Electrocoagulation

The book is a concise presentation of different applications of the electrocoagulation treatment process involved in the removal of both organic and inorganic contaminants from drinking and industrial wastewater. The fundamental concept concerning the present status of water and wastewater treatment is elaborately explained.

Features:

- Addresses the hybrid electrocoagulation approaches established for the treatment of toxic refractory contaminants.
- Discusses the current status and the limitations of the electrocoagulation process, as well as the advancement in the field of modeling and simulation of electrocoagulation reactors.
- Describes the economic and environmental feasibility of the process, along with future suggestions/recommendations on the electrocoagulation process.
- Discusses the existing loopholes and explains the areas in which improvement can be made to make the water treatment process more feasible.
- Explores the experimental studies conducted, along with their associated challenges and possible solutions, thereby providing comprehensive knowledge to the readers in the field of water and wastewater treatment, starting from the basics and gradually progressing toward the advanced level of the subject.

As well as undergraduate and postgraduate students, this book is strongly recommended for people concerned with the research and development sectors of water reuse, environmental restoration, and industrial wastewater treatment.

Electrocoagulation
Fundamentals and Applications in Water and Wastewater Treatment

Mihir Kumar Purkait, Pranjal Pratim Das
and
Mukesh Bharti

CRC Press
Taylor & Francis Group
Boca Raton London New York

CRC Press is an imprint of the
Taylor & Francis Group, an **informa** business

Designed cover image: Getty Images

First edition published 2025
by CRC Press
2385 NW Executive Center Drive, Suite 320, Boca Raton FL 33431

and by CRC Press
4 Park Square, Milton Park, Abingdon, Oxon, OX14 4RN

CRC Press is an imprint of Taylor & Francis Group, LLC

ISBN: 978-1-032-79650-5 (hbk)
ISBN: 978-1-032-81404-9 (pbk)
ISBN: 978-1-003-49967-1 (ebk)

DOI: 10.1201/9781003499671

Typeset in Times
by Apex CoVantage, LLC

This book is dedicated to all the researchers, engineers, and working professionals whose groundbreaking efforts have inspired solutions towards global water challenges and achieving a sustainable future. This work is a testament to your efforts and a small contribution to our shared vision of unwavering commitment to clean water and environmental sustainability.

Contents

Preface

Water is essential for human survival, yet its scarcity and contamination are among today's most pressing global challenges. Industrialization, population growth, and climate change have intensified the demand for clean water and increased the complexity of water contamination issues. This book offers an in-depth exploration of electrocoagulation, an efficient and environmentally friendly water treatment technology. It comprehensively examines various water treatment methods, including physical, chemical, and biological approaches, which have long been foundational in water and wastewater treatment. In addition, this book reviews the current regulatory discharge criteria for drinking water and industrial wastewater. The book also covers the fundamentals of electrochemistry, including electrochemical cells, reference electrodes, and reaction rate variables, which are central to understanding electrocoagulation. It also explains pollutant removal mechanisms by generating metal ions from sacrificial iron or aluminum electrodes under an electric current, producing hydroxides that coagulate and adsorb contaminants for effective removal. Key parameters such as electrode material, current density, initial pH, and supporting electrolytes—factors significantly influencing electrocoagulation performance—are thoroughly discussed.

Furthermore, this book describes optimization strategies for reactor performance, including modeling fluid dynamics, mass transfer, and reaction kinetics. It addresses challenges in scaling up electrocoagulation systems and provides case studies demonstrating the practical application of these models. The application of the EC process to remove various contaminants commonly found in drinking water (e.g., arsenic, fluoride, iron, and microplastics) and industrial wastewater (e.g., heavy metals, dyes, and organic compounds) is comprehensively reviewed. Real-world case studies are presented, focusing on the steel, textile, and pharmaceutical industries. The book also discusses sludge management and valorization strategies, which are crucial for the sustainability of electrocoagulation systems. An in-depth analysis of hybrid systems integrated with electrocoagulation—including EC-membrane filtration, EC-oxidation, and EC-adsorption—is provided, comparing the advantages of these systems with stand-alone processes. Additional case studies highlight the treatment of various contaminants in industrial wastewater (including wine, textile, steel, and petroleum industries) and drinking water from diverse regions around the world. These case studies offer practical insights into the challenges and successes of implementing EC technology on a large scale.

In addition, the book assesses the life cycle of the EC system, examining its carbon footprint, energy demands, and the quality of treated water. Economic aspects of scaling up the EC process are also considered, including a cost–benefit analysis that highlights the long-term advantages of the technology. The potential integration of the EC process with hydrogen production systems is explored, presenting an opportunity for dual benefits in water treatment and energy generation. This book aims to serve as a comprehensive resource for students, researchers, and professionals involved in water and wastewater treatment. By bridging theoretical concepts with practical applications, it provides readers with an extensive understanding of the electrocoagulation process and its potential to transform water treatment technologies in the coming years.

Dr. Mihir Kumar Purkait
Dr. Pranjal Pratim Das
Mr. Mukesh Bharti

Author Biographies

Dr. Mihir Kumar Purkait is a professor in the Department of Chemical Engineering at the Indian Institute of Technology Guwahati, Assam, India. His current research activities focus on four distinct areas: (i) smart materials, (ii) waste to energy, (iii) process intensification, and (iv) wastewater treatment. In each of these areas, his goal is to synthesize stimuli-responsive materials and to develop a more fundamental understanding of the factors governing the performance of chemical and biochemical processes. He has more than 20 years of experience in academics and research and has published more than 300 papers in different reputed journals (Citation: >16,500, h-index = 75, i-10 index = 193). He has 12 patents and completed 43 sponsored and consultancy projects from various funding agencies.

Dr. Pranjal Pratim Das is a technical associate at the National Jal Jeevan Mission (NJJM) under the Ministry of Jal Shakti, Govt. of India. He has completed his PhD from the Department of Chemical Engineering, Indian Institute of Technology Guwahati, Assam, India. He received his M.Tech and B.Tech in Food Engineering and Technology from Tezpur (Central) University, Assam, India. His research work is purely dedicated to industrial wastewater treatment via electrochemical and advanced oxidation techniques. He has extensively worked on the application of the hybrid ozone–electrocoagulation process to treat heavy metals and cyanide-contaminated effluents from different unit operations of steel industry. He has authored several scientific book publications, research/review articles, and book chapters in various reputed international journals on water and wastewater treatment. He has fabricated and demonstrated many lab-scale modules for the green energy generation from sewage wastewaters. He has also worked on the treatment of ground and surface waters and has delivered many pilot plant setups to several water treatment facilities across the state of Assam (India) for the supply of safe drinking water.

Mr. Mukesh Bharti received his Diploma (2010) in Chemical Engineering from P.M.V. Polytechnic, Mathura, India. He joined Shri Ram Fluorochemicals Ltd., Bhiwadi, Rajasthan, as a junior engineer and worked for two years in the Effluent Treatment Plant. He then received his B.Tech in chemical engineering from Harcourt Butler Technological Institute (HBTI) Kanpur, Uttar Pradesh, India. Later he joined Frontier Pulp & Paper Mill industry and worked in its Effluent Treatment Plant for four years. After that he obtained his M.Tech and is currently a research scholar in the Department of Chemical Engineering at the Indian Institute of Technology Guwahati, Assam, India. His research is focused on water treatment via magnetic electrocoagulation and membrane processes.

Acknowledgements

The successful completion of this book has been a long scholastic endeavour embedded with affection, trust and support from some of the most valuable people and communities of our lives. We would like to extend our sincere appreciation to our parents, family, friends, and well-wishers. Without their passionate assistance and active participation, this book would not have taken shape. Further, the honour is elevated manifold because of the opportunity bestowed upon us by the following esteemed institutions: Indian Institute of Technology Guwahati (IITG), Assam, India, and National Jal Jeevan Mission (NJJM), India. We express our heartfelt gratitude to both these institutions.

1 An Overview of Water and Wastewater Treatment

1.1 INTRODUCTION

Water is the essence of life on our planet and its significance cannot be overstated as it plays a pivotal role in ecosystems, human health, agriculture, industry, and countless other aspects of society. However, as the global population grows and human activities intensify, the demand for clean, safe water increases while the quality of available water sources deteriorates. This necessitates the implementation of effective water and wastewater treatment processes to ensure the provision of potable water and the protection of the environment [1].

The availability of clean water is essential for drinking, sanitation, hygiene, and food production. However, various natural and anthropogenic factors threaten the quality of water sources, including pollution from industrial discharge, agricultural runoff, sewage, and chemical contaminants. Without adequate treatment, contaminated water poses significant risks to human health and the environment, leading to waterborne diseases, ecosystem degradation, and economic losses. Therefore, treatment processes are indispensable for removing contaminants, reducing pollution, and minimizing the ecological footprint of human activities. Water and wastewater treatment processes are designed to remove impurities, pathogens, and pollutants from raw water and sewage, rendering them safe for various applications or suitable for discharge into the environment. These treatment methods rely on physical, chemical, and biological mechanisms to achieve purification and disinfection. Water and wastewater treatment are essential processes that ensure the safety and sustainability of water resources for human consumption and environmental health. This chapter explores the key aspects of water and wastewater treatment, including their importance, the processes involved, and emerging technologies [2].

Water treatment is the method of eliminating contaminants from water to make it suitable for various purposes, such as drinking, irrigation, and industrial processes. Wastewater treatment involves the removal of pollutants from used water before it is discharged back into the environment. Both water and wastewater treatment play crucial roles in maintaining public health, protecting ecosystems, and conserving water resources. The primary goal of water treatment is to produce clean and safe drinking water. The treatment process varies depending on the type of pollutants present and the desired water quality for discharge. Common steps in wastewater treatment include preliminary treatment (screening and grit removal), primary treatment (settling to remove solids), secondary treatment (biological processes to

DOI: 10.1201/9781003499671-1

remove organic matter), and tertiary treatment (additional filtration or disinfection). Additionally, emerging technologies are revolutionizing water and wastewater treatment. Advanced oxidation processes, nanotechnology, and membrane filtration are examples of innovative approaches that improve treatment efficiency and produce higher-quality water [3].

This chapter primarily focuses on the water purification cycle, basic water chemistry, water and wastewater treatment processes and pollutants discharge standards. A detailed analysis has also been done on the treatment processes, such as physical, chemical and biochemical for treating water and wastewater. Additionally, a thorough discussion is done on the disposal of residues and biosolids derived from sewage sludge. Residues and biosolids may contain various contaminants, including heavy metals, organic pollutants, pathogens, and emerging contaminants like pharmaceuticals and personal care products. If not properly managed, these contaminants can leach into soil and water, posing risks to human health and the environment.

1.2 WATER PURIFICATION CYCLE

The water purification cycle is a natural process that involves the movement and transformation of water through various stages, resulting in the purification and distribution of freshwater across the Earth's surface. This cycle, also known as the hydrological cycle or water cycle, plays a fundamental role in maintaining the availability of clean water for ecosystems, human utilization, industrial processes, and agricultural use. The water purification cycle is illustrated in Figure 1.1. There

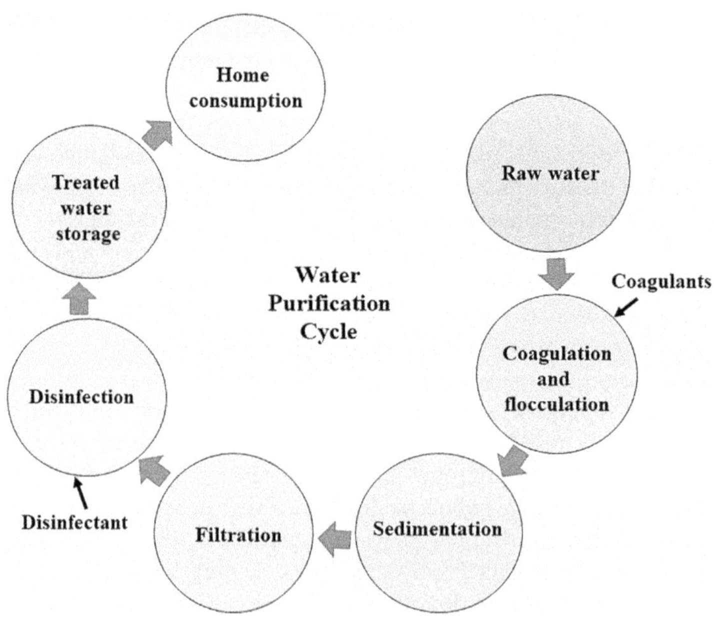

FIGURE 1.1 Water purification cycle.

are five steps in the treatment of water that is intended for drinking or cooking: screening, coagulation, sedimentation, filtration, and disinfection. Large contaminants like fish, leaves, and twigs are removed from the water using coarse or fine screens during the screening process, which is the first stage in the water treatment process. This straightforward yet efficient procedure aids in preventing harm to the treatment facility. Impurities (fine particles) are further removed from the water by the processes of coagulation and sedimentation. During this procedure, coagulants like ferric chloride and aluminum sulfate are added to the water to create flocs, which draw in dirt particles and finally cause them to settle to the bottom of the unit (sediment) due to their heavy weight. Additionally, this gets rid of any viruses and germs that are present in the water.

The following stage is the filtration process, which, as its name implies, uses a variety of filters, including sand and gravel beds, to assist remove sediment—solid particles—from the clean water in the unit. Following this, the last disinfection step is executed. The water is next treated with chemicals like chlorine or chlorine dioxide to get rid of any bacteria or germs, after which it is securely kept in tanks before being piped into homes. Comprehending the water purification cycle is essential in the effective management of water resources, mitigating water-related challenges, and ensuring sustainable water management practices. This chapter delves into the different stages of the water purification cycle, their mechanisms, and their significance in maintaining the global water balance [4].

1.3 BASIC WATER CHEMISTRY

It is beneficial to know the fundamentals of water chemistry and the ions that are frequently present in water. This chapter covers water quality guidelines, some typical compounds found in commercial and industrial water supply systems, and solubility concepts. It may assist the reader in understanding the various kinds of pollutants and how they affect water systems. Since water can dissolve any substance to some degree, it is known as the universal solvent. Its structure is a key to this unique capability. Both the oxygen and hydrogen atoms that make up the water molecule have little positive and negative charges with each other, respectively. Water is a solvent that dissolves all polar and ionic materials because of its polarity. Insoluble in water are non-polar substances like hydrocarbons. The surrounding environment does not contain clean water as a consequence. The taste, smell, feel, color, clarity, and scent are all imparted by contaminants, which are always present. Four categories of contaminants can be identified in water and wastewater: dissolved, physical, microbiological, and radioactive. Natural waters' characteristics vary depending on where they come from. Surface waterways include rivers and lakes, whereas groundwater includes wells and springs. As contaminants penetrate through the layers of rock, groundwater accumulates them. The hydrological cycle is made up of surface and groundwater, both of which are essential to maintaining ecosystems and providing for human water needs [5]. The term "groundwater" describes the water that is kept below the surface of the Earth in aquifers, which are sedimentary or rock layers that are porous and can hold and transfer water. This water originates from precipitation, infiltration, and surface water runoff that percolates downward over time. Groundwater serves as a natural reservoir, providing a reliable source of freshwater for drinking, agriculture, and industry and maintaining stream flow during dry periods. In contrast, surface water includes rivers, lakes, streams, and reservoirs that are visible above the Earth's surface. It primarily originates from precipitation runoff, snowmelt, and direct rainfall onto water bodies. Surface water plays a crucial role in supporting aquatic habitats, providing recreational opportunities, and serving as a source of water for various human activities like irrigation, navigation, and hydropower generation.

1.3.1 Miscibility and Solubility

Miscibility refers to the ability of substances to mix uniformly at the molecular level, forming a homogeneous solution. In the context of water treatment, miscibility plays a crucial role in the dissolution and dispersion of chemicals, additives, and treatment agents in water to achieve desired treatment objectives. Water-soluble substances, such as salts, acids, bases, and organic compounds, exhibit varying degrees of miscibility with water, depending on their chemical properties and molecular structures.

Similarly, the solubility characteristics of substances in water are essential for designing and implementing effective water treatment processes, including chemical dosing, disinfection, coagulation, flocculation, and pH adjustment. By selecting appropriate chemicals and additives with high solubility and miscibility in water, water treatment operators can optimize treatment efficiency, enhance process performance, and achieve desired water quality standards.

1.3.2 Suspension, Sediment, and Particles

Suspension, sediment, and particles play critical roles in the dynamics of water and wastewater treatment processes, influencing the quality, clarity, and stability of aqueous systems. Understanding these concepts is essential for effective management and control of suspended solids in water treatment plants, industrial processes, and environmental remediation efforts.

1.3.2.1 Suspension

Suspension refers to a heterogeneous mixture in which solid particles are dispersed or suspended in a liquid medium, such as water or wastewater. These suspended particles are typically insoluble or only partially soluble in the liquid phase, giving rise to a colloidal system characterized by the continuous movement of particles within the fluid. The sources of suspended solids are described below:

- **Natural Sources:** Suspended solids in natural waters originate from erosion, weathering, sedimentation, biological activity, and atmospheric deposition. Soil erosion, river runoff, and coastal erosion contribute significant amounts of suspended sediment to aquatic ecosystems.

- **Anthropogenic Sources:** Human activities, such as urbanization, agriculture, mining, construction, and industrial processes, introduce suspended solids into water bodies through storm-water runoff, wastewater discharges, and pollutant emissions. Industrial effluents, sewage discharges, and dredging operations can contain elevated concentrations of suspended particles.

1.3.2.2 Sediment

Sediment refers to the solid material that settles or accumulates at the bottom of a liquid or on surfaces within a water body, forming sedimentary deposits. Sedimentation is a natural process driven by gravity, hydrodynamics, and particle interactions, shaping aquatic environments and influencing water quality, habitat stability, and geomorphology.

1.3.2.3 Particles

Particles refer to discrete solid entities or entities of matter, ranging in size from subatomic particles to macroscopic particles visible to the naked eye. In water and wastewater treatment, particles include suspended solids, colloids, flocs, aggregates, and particulate matter that influence the clarity, quality, and properties of aqueous systems. The various types of particles are described below:

- **Suspended Solids:** Suspended solids are solid particles dispersed or suspended in water or wastewater, encompassing a range of sizes, shapes, compositions, and origins. Suspended solids can include clay, silt, organic matter, microorganisms, algae, and synthetic particles from anthropogenic sources.
- **Colloids:** Colloidal particles are small particles with diameters between 1 nanometer and 1 μm, exhibiting Brownian motion and electrostatic repulsion in aqueous solutions. Colloids can be stable or unstable, depending on factors such as particle size, surface charge, and physicochemical conditions.
- **Flocs:** Flocs are agglomerates or clusters of suspended particles formed through flocculation or coagulation processes in water treatment. Flocs consist of entangled particles held together by physical or chemical forces, enhancing their settleability and removal efficiency in sedimentation or filtration.
- **Aggregates:** Aggregates are larger particles formed by the aggregation or association of smaller particles or colloids in water. Aggregates can range in size from micrometers to millimeters and may include loose aggregates, micro aggregates, and macro aggregates, depending on the aggregation mechanism and stability [7].

1.3.3 Ions and Emulsions

1.3.3.1 Ions

Ions are electrically charged atoms or molecules that result from the gain or loss of electrons. In water and wastewater, ions can be either positively charged (cations) or negatively charged (anions) depending on whether they have lost or gained electrons, respectively. Certain ions, such as nitrate (NO_3^-), phosphate (PO_4^{3-}), and sulfate (SO_4^{2-}), serve as nutrients for microbial growth. While essential in moderate concentrations, excessive levels of nutrients can lead to eutrophication in receiving waters, causing algal blooms and oxygen depletion. In addition, ions like sodium (Na^+), chloride (Cl^-), and calcium (Ca^{2+}) contribute to the salinity of water. High salinity levels in water sources, often due to natural processes or anthropogenic activities like irrigation or industrial discharges, can affect water quality and freshwater ecosystems. Moreover, hardness ions, primarily calcium (Ca^{2+}) and magnesium (Mg^{2+}), contribute to water hardness, affecting its suitability for domestic and industrial use [8].

1.3.3.2 Emulsions

An emulsion is a mixture of two or more immiscible substances, typically consisting of one liquid dispersed in another in the form of tiny droplets. In water and wastewater treatment, emulsions often involve the dispersion of oils, greases, or organic compounds in water. Emulsified oils, greases, and organic compounds can enter water bodies through various sources, including industrial discharges, urban runoff, and wastewater effluents. These contaminants can degrade water quality, harm aquatic ecosystems, and pose risks to human health. In addition, emulsions can hinder the effectiveness of conventional water and wastewater treatment processes by forming stable colloidal suspensions that are resistant to settling, filtration, and biological degradation. Addressing emulsified contaminants is essential for achieving desired treatment outcomes. However, managing both (ions and emulsions) effectively is essential for ensuring the efficiency, reliability, and environmental sustainability of water and wastewater treatment systems [9].

1.4 WATER AND WASTEWATER TREATMENT PROCESSES

Water treatment processes encompass a variety of methods designed to remove impurities, contaminants, and pathogens from raw water to make it safe for consumption, industrial use, or environmental discharge. These processes can be categorized into several types based on their mechanisms and objectives. Here are the main types of water treatment processes:

1.4.1 Physical Treatment Processes

1.4.1.1 Screening

Screening involves the removal of large debris, such as leaves, sticks, and trash, from raw water using screens or grates. This process prevents damage to downstream equipment and ensures the efficiency of subsequent treatment processes.

1.4.1.2 Sedimentation

Sedimentation, also known as settling, allows suspended particles and floc to settle out of water under the influence of gravity. In sedimentation tanks or clarifiers, the velocity of water flow is reduced, allowing particles to settle at the bottom of the tank, forming a sludge layer that can be removed.

1.4.1.3 Filtration

Filtration is a mechanical process that involves passing water through porous media, such as sand, gravel, or membranes, to remove suspended particles, sediment, and microbial contaminants. Filtration can be achieved through rapid sand filtration, slow sand filtration, or membrane filtration [10].

1.4.2 Chemical Treatment Processes

1.4.2.1 Coagulation and Flocculation

Coagulation and flocculation are chemical processes used to remove suspended particles, colloids, and organic matter from water. Coagulants, such as aluminum sulfate (alum) or ferric chloride, are added to water to neutralize charges and promote particle aggregation. Flocculants, such as polymers, are then added to enhance the formation of larger floc particles, which can be easily removed by sedimentation or filtration.

1.4.2.2 pH Adjustment

pH adjustment is often required to optimize the effectiveness of water treatment processes and ensure compliance with regulatory standards. Adding chemicals, such as lime (calcium hydroxide) or soda ash (sodium carbonate), can raise or lower the pH of water to the desired range, enhancing the removal of contaminants and improving water quality.

1.4.3 Biological Treatment Processes

Biological treatment processes are a natural and efficient way to treat both water and wastewater. They utilize microorganisms, such as bacteria, fungi, and protozoa, to break down organic contaminants and pollutants. These processes are widely used in municipal and industrial wastewater treatment plants (WWTPs) as well as for drinking water treatment in some cases.

There are two main types of biological treatment processes: aerobic and anaerobic process.

1.4.3.1 Aerobic Process

Aerobic treatment processes occur in the presence of oxygen. Microorganisms use oxygen to break down organic matter into simpler compounds, such as carbon dioxide and water. Some common examples of aerobic treatment processes include the following:

- **Activated sludge:** Activated sludge is a powerful secondary treatment procedure that uses hungry microbes to clean wastewater even further. The primary treatment removes large solids, but there are still tiny particles and dissolved organic matter left behind. In activated sludge tanks, air is

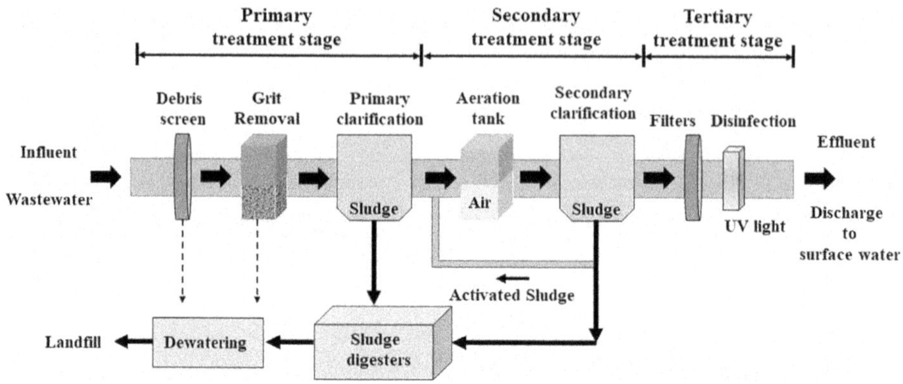

FIGURE 1.2 Activated sludge treatment process.

pumped into the wastewater to create a bubbly mix. This allows a brown, floc-like material called "activated sludge" to grow. This "floc" is made up of countless bacteria and other organisms that feed on the organic waste in the water. The complex organic matter in the water acts like a giant buffet for microbes. Some bacteria break down the large, complex molecules into simpler ones, like leftovers they do not eat themselves. Other bacteria gobble up these leftovers, breaking them down even further until nothing is left that the microbes can use as food. Agitation (mixing) is also important to prevent the sludge from settling at the bottom and ensure good contact between the microbes and the wastewater [11]. The whole sludge treatment process is shown in Figure 1.2.

The wastewater treatment process begins with the collection of effluent from industries through an underground drainage system. This effluent is then directed to treatment plants, where it undergoes multiple treatment stages including primary, secondary, and tertiary treatments. Initially, before treatment, the effluent goes through stages such as screening, which is a unit operation used for odor control and prevention of sticking and clogging. This screening process eliminates materials like plastics, paper, and other debris, while chemicals are used for neutralization. Coarse screens are used to remove large solids and debris, while fine screens are employed to eliminate substances that could cause operating and maintenance issues in downstream processing [12].

1.4.3.2 Primary Treatment

The first stage of water treatment, called primary clarification, focuses on removing large particles and organic matter. Big tanks are used to slow down the water, allowing heavy solids to sink and lighter materials like fats and oils to float. Scrapers skim these off the surface, while grit and debris get caught at the bottom. This stage uses equipment like settling tanks (also called clarifiers), septic tanks, and some specialized systems. To make removal easier, grinders are sometimes used to break down large objects into smaller pieces. Additionally, some plants include special tanks to capture even finer particles by allowing the water to settle for longer periods. Overall, primary treatment aims to clean the water enough for further purification steps in the next stage, secondary treatment.

1.4.3.3 Preliminary Treatment: Screening and Grit Removal

Wastewater often contains rocks, sand, and other debris that could damage equipment during treatment. Screening removes these unwanted materials. A screen is like a colander with evenly sized holes that trap large objects like rags and food scraps. The holes can be shaped like circles or slots, and smaller holes catch more debris. Grit, specifically, refers to the heavy inorganic materials like sand and eggshells that sink in wastewater. It is heavier than the organic waste that decomposes naturally. Removing grit protects equipment from getting worn down. Grit chambers are designed to capture just the heavy stuff, and ideally, the grit that comes out is mostly clean sand and rocks suitable for landfill use. If there is a lot of organic material mixed in it, the grit might need to be buried or used as fertilizer (depending on regulations) [13].

1.4.3.4 Sedimentation

Sedimentation is a natural process where gravity helps remove heavy things from wastewater. Imagine letting a muddy puddle settle—the dirt eventually sinks to the bottom. In wastewater treatment plants, this happens in large tanks that slow down the flow, allowing things like sand, clay, and clumps of organic matter (flocs) to sink out as sludge. This is a big advantage because it reduces the need for additional chemicals to clump particles together before settling (coagulation and flocculation). The sedimentation process can also be used after these chemicals are added to improve how well the clumps settle out. Special tanks are used for wastewater sedimentation. These come in different shapes such as long channels (horizontal flow), layered compartments (multi-layer), or circular basins (radial flow). Some tanks even use a "floc blanket," a layer of already settled material that helps trap even finer particles as the water flows through. It is important to note that sedimentation is just one part of primary treatment. Other techniques like spinning the water to separate solids (centrifugal separation), adjusting acidity (neutralization), or using air bubbles to float out lighter materials (dissolved air flotation) might also be used in the initial stage [14].

1.4.3.5 Secondary Treatment

Secondary treatment uses tiny organisms (microbes) to eat up the organic matter in the water. These microbes act like a living filter, breaking down the waste and leaving behind harmless by-products. However, this biological treatment can occur in two main ways:

- **Attached growth:** In this method, the microbes are "attached" to a solid surface like rocks or plastic carriers inside a tank. The wastewater flows past these surfaces, and the microbes munch on the organic matter in the water. For instance, rotating biological contactors and trickling filters.

- **Suspended growth:** This approach uses freely moving microbes suspended throughout the water in large tanks. These microbes, often called "activated sludge," are constantly fed air to keep them happy and hungry. Activated sludge is a common and effective secondary treatment process.

No matter which method is used, secondary treatment removes leftover organic matter and suspended solids from primary treatment. This significantly cleans the water and prepares it for the final polishing stage, tertiary treatment [15].

1.4.3.6 Tertiary Treatment

Tertiary wastewater treatment becomes necessary to eliminate persistent contaminants, nutrients, and pathogens after the initial treatment stages. This step is crucial to meet the quality standards for reuse and to render the treated water suitable for land applications or direct discharge into water bodies like ponds, lakes, and rivers. Tertiary treatment encompasses processes such as nutrient removal, disinfection, ion exchange, membrane processes, and filtration. Chlorine disinfection and these methods prove effective in eliminating emerging environmental pollutants such as antibiotic resistance genes in wastewater effluent [16].

1.4.3.7 Disinfection (UV/Chlorination)

The term "disinfection" describes the procedure used to kill, eliminate, or eradicate bacteria. The purpose of adding these disinfectants to the wastewater treatment process is to fight off the bacteria found in pollutants. UV light or chlorine can be used to do this. Chlorine is normally a greenish-yellow gas, but at high pressure, it becomes liquid. When added to water, it successfully eradicates almost all harmful germs. Tertiary treatment also makes substantial use of ultraviolet, or UV, radiation. UV radiation regulates the capacity for regrowth within the transmission system and disinfects by changing the biological components of bacteria, especially by breaking the chemical bonds in DNA, RNA, and proteins. In the tertiary treatment stage, methods such as electrodialysis, reverse osmosis, and granular or activated carbon are used [17].

- **Trickling Filters:** In this process, wastewater is sprayed over a bed of rocks or other media that is covered with a biofilm of microorganisms. The microorganisms break down the organic matter in the wastewater as it passes through the biofilm. These filters are made of a cylindrical tank packed with a high-specific surface area material, including gravel, rocks, broken PVC bottles, or specific pre-made plastic filter media. In addition to its large surface area, the medium surface forms a thin biofilm on the surface on which organisms live. It oxidizes the organic load in wastewater, generating water, carbon dioxide, and new biomass. The pretreated wastewater is trickled over the filter, usually with a sprinkler system (as shown in Figure 1.3) that rotates, offering the filter media opportunities to be dosed and exposed to air in cycles. In addition, techniques such as extended aeration, disc contractors, aerated lagoons, contacting media, and high-rate activated sludge are applied during the secondary treatment [18].

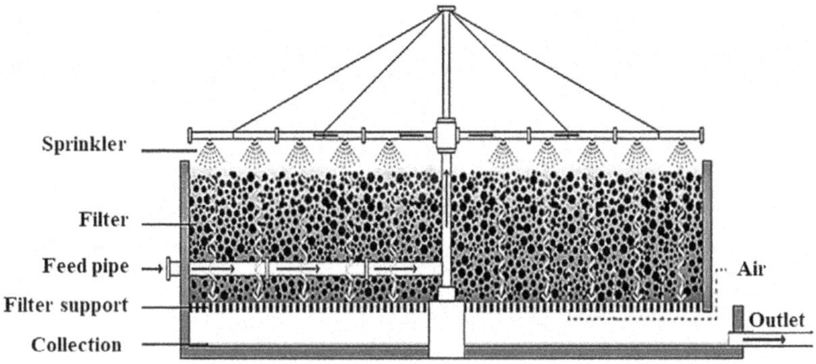

FIGURE 1.3 Biological trickling filter system for wastewater treatment.

- **Constructed Wetlands:** These are human-made wetlands that are designed to treat wastewater. They are planted with aquatic plants, which help to remove pollutants from the water through a combination of biological, physical, and chemical processes [19].

1.4.3.8 Anaerobic Process

Anaerobic treatment processes occur in the absence of oxygen. Microorganisms break down organic matter into methane, carbon dioxide, and other by-products. Anaerobic treatment is often used to treat industrial wastewater that contains high levels of organic matter. It is also used in some sludge digestion processes to reduce the volume of sludge that needs to be disposed of. Some common examples of anaerobic treatment processes include the following:

- **Anaerobic lagoons:** These are large ponds or lagoons that are used to treat wastewater anaerobically, as shown in Figure 1.4. The wastewater is mixed slowly in the lagoon, which allows the anaerobic microorganisms to break down organic matter.

This process significantly reduces the amount of organic material in the wastewater, though it can create unpleasant odors. Anaerobic ponds are often the first stage in a multi-pond treatment system, with subsequent ponds reintroducing oxygen for further breakdown and purification.

- **Anaerobic digesters:** These are closed tanks that are used to treat wastewater or sludge anaerobically. The digesters are heated to maintain a warm temperature, which is optimal for the growth of anaerobic microorganisms. The methane gas produced by the digesters can be used as a renewable energy source.

In addition, biological treatment processes are a cost-effective and sustainable way to treat water and wastewater. They are also becoming increasingly popular as regulations on wastewater discharge become more stringent [20].

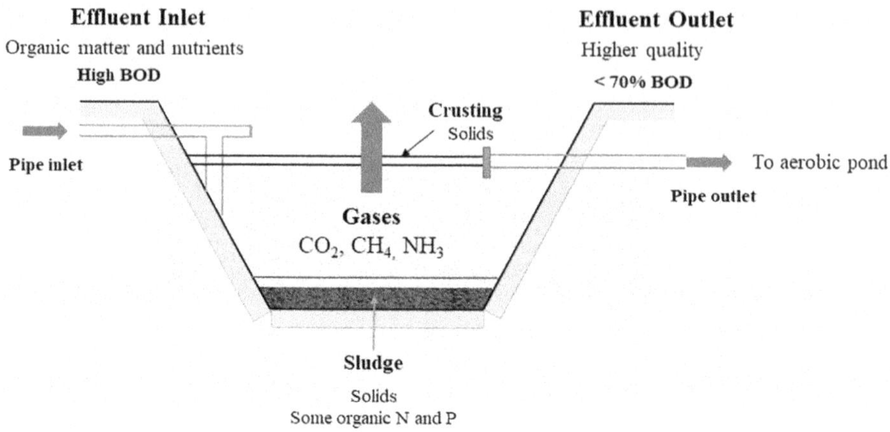

FIGURE 1.4 Schematic diagram of anaerobic pond.

1.5 REGULATORY DISCHARGE STANDARDS

Regulatory discharge standards for water and wastewater play a crucial role in safeguarding our environment. These standards are established to ensure that industrial and municipal discharges meet specific criteria, minimizing the impact on water bodies.

1.5.1 Drinking Water

1.5.1.1 Drinking Water Guidelines

Regulatory discharge standards for drinking water vary by country and region and are typically set by governmental or regulatory agencies responsible for public health and environmental protection. These standards establish maximum allowable concentrations of various contaminants in drinking water to ensure that it is safe for consumption. Here are some key aspects of regulatory discharge standards for drinking water.

1.5.1.2 Contaminants Covered

Regulatory standards typically address a wide range of contaminants that may be present in drinking water, including microbial pathogens (such as bacteria, viruses, and protozoa), chemical pollutants (such as heavy metals, pesticides, organic compounds, and disinfection by-products), and physical parameters (such as turbidity and color).

1.5.1.3 Maximum Contaminant Levels (MCLs)

MCLs represent the maximum allowable concentrations of contaminants in drinking water as specified by regulations. These levels are based on scientific assessments of the potential health risks associated with exposure to specific contaminants over a lifetime of consumption.

1.5.1.4 Monitoring and Compliance

Regulatory agencies typically require water utilities to regularly monitor the quality of their drinking water supplies to ensure compliance with established standards. This involves sampling and testing water samples for various contaminants according to prescribed methods and frequencies.

1.5.1.5 Treatment Requirements

In many cases, water treatment facilities are required to implement specific treatment processes to achieve compliance with regulatory standards. These treatments may include disinfection (e.g., chlorination and ozonation), filtration (e.g., activated carbon and membrane filtration), and other advanced treatment technologies to remove or reduce contaminants to acceptable levels.

1.5.1.6 Public Notification

Regulatory agencies often mandate that water utilities inform the public about the quality of their drinking water, including any violations of regulatory standards. Public notification requirements ensure transparency and allow consumers to make informed decisions about their water consumption.

1.5.1.7 Updates and Revisions

Regulatory discharge standards for drinking water are subject to periodic review and updates to reflect advances in scientific knowledge, changes in public health priorities, emerging contaminants, and evolving treatment technologies. Regulatory agencies may revise the MCLs and other requirements based on these considerations. Examples of regulatory agencies responsible for setting and enforcing drinking water standards include the United States Environmental Protection Agency (EPA) with its Safe Drinking Water Act (SDWA), the European Union with its Drinking Water Directive, and various national and local agencies in other countries.

However, water utilities and consumers need to stay informed about applicable regulatory standards to ensure the safety and quality of drinking water supplies.

1.5.1.8 Effluent Guidelines

Developed by the Environmental Protection Agency (EPA), these guidelines set national wastewater discharge standards for various industries. They are technology-based, aiming to achieve the maximum pollutant reduction that is economically feasible for each sector. Effluent guidelines address both surface water discharges and municipal sewage treatment plants.

(a) Parameters and Limits

- Different countries have their regulatory bodies (such as the Central Pollution Control Board (CPCB) in India) that define specific parameters and limits.
- These standards cover various pollutants, including chemical oxygen demand (COD), total suspended solids (TSS), pH, heavy metals, and more.

- Compliance with these limits ensures that discharged wastewater does not harm aquatic ecosystems or public health.

(b) Monitoring and Enforcement

- Regular monitoring ensures that industries adhere to the established standards.
- Violations can result in penalties, shutdowns, or legal actions.
- Effective enforcement mechanisms are essential to maintain water quality.

These standards are vital for sustainable water management and the well-being of our planet.

The World Health Organization (WHO) develops international norms on human health and water quality in the form of guidelines that serve as the foundation for international standards and regulations in both developed and developing nations. The purity of drinking water is one of the most important aspects of the environment that influences health. Water-related disease prevention and control start with assuring the safety of drinking water. The WHO Expert Advisory Committee consults with national authorities, international agencies, and WHO countries, including India, to prepare the guidelines. This extensive global consultative process results in the guidelines being drafted by the WHO. The drinking water standards are given in Table 1.1 as per the WHO guidelines.

1.5.2 Industrial Wastewater

Industrial wastewater refers to the water generated by various industries during their production processes. Unlike domestic wastewater (coming from homes), it can be significantly more complex and contain various pollutants depending on the specific industry. Chemical industries have a significant environmental impact compared to other industries. The wastewater discharge is typically rich in organic and inorganic pollutants, often with toxic properties. Many of these effluents are found to be mutagenic, carcinogenic, and non-biodegradable [22]. Various sources of industrial wastewater are described below.

1.5.2.1 Iron and Steel Industry

The wastewater produced from the iron and steel industry contains hazardous substances such as ammonia, phenol, and cyanide which pose a threat to the water bodies where they are discharged. The raw wastewater produced by the steel industry has several detrimental effects, including toxicity to aquatic life and a decrease in dissolved oxygen (DO) levels due to the presence of suspended solids like ammonia and phenol [23]. These substances can raise the water's pH, contributing to its toxicity. Moreover, the release of decomposable organic matter into water bodies results in soil and bacteria using this matter as a carbon source, leading to a reduction in DO levels in the water [24].

1.5.2.2 Food Industry

The wastewater discharged by the food industry contributes to environmental pollution because of its prominent levels of COD and BOD. Unlike many other industries, the food sector necessitates a substantial volume of water, which is utilized

TABLE 1.1
Drinking water standards (Reproduced with permission from Cotruvo et al. (1988) @Elsevier [21]).

S. No.	Characteristics	Maximum acceptable limit	Permissible limit
(a) Organoleptic and Physical Parameters			
1.	Color, Hazen units	5	15
2.	Odor	Agreeable	Agreeable
3.	pH Value	6.5–8.5	No relaxation
4.	Taste	Agreeable	Agreeable
5.	Turbidity, NTU	1	5
6.	Total dissolved solids (TDS), mg/L	500	2000
(b) General Parameters			
1.	Aluminum (Al)	0.03	0.2
2.	Ammonia	0.5	No relaxation
3.	Anionic detergents (MBAS)	0.2	1.0
4.	Barium (Ba)	0.7	No relaxation
5.	Boron (B)	0.5	1.0
6.	Calcium (Ca)	75	200
7.	Chloramines (Cl_2)	4.0	No relaxation
8.	Chloride (Cl)	250	1000
9.	Copper (Cu)	0.05	1.5
10.	Fluoride (F)	1.0	1.5
11.	Free residual chlorine	0.2	1.0
12.	Iron (Fe)	0.3	No relaxation
13.	Magnesium (Mg)	30	100
14.	Manganese (Mn)	0.1	0.3
15.	Mineral oil	0.5	No relaxation
16.	Nitrate (NO_3)	45	No relaxation
17.	Phenolic compounds	0.001	0.002
18.	Selenium (Se)	0.01	No relaxation
19.	Silver (Ag)	0.1	No relaxation
20.	Sulfate (SO_4)	200	400
21.	Sulfide (H_2S)	0.05	No relaxation
22.	Total alkalinity ($CaCO_3$)	200	600
23.	Total hardness ($CaCO_3$)	200	600
24.	Zinc (Zn)	5	15

extensively in various plant processes. Noukeu NA et al. (2016) observed variations in suspended solids, BOD/COD ratios, and total solids within the effluents of these industries, attributed to the diverse additives employed in different food products. Particularly, the chocolate industry stands out as one of the most environmentally taxing in terms of water consumption [25].

1.5.2.3 Dairy Industry

Typically, wastewater from the dairy sector is characterized by a significant presence of solid organic compounds. This industry primarily deals with substantial volumes of milk, which contains a range of constituents such as casein and inorganic salts. The effluent's quality largely depends on the volume of milk processed and the specific products manufactured. Dairy effluent treatment primarily involves physicochemical approaches such as coagulation/flocculation, chemical precipitation, and membrane processes, which can achieve efficiency rates of up to 98% [26].

1.5.2.4 Leather Industry

The leather industry stands out as one of the rapidly expanding and highly polluting sectors worldwide, contributing significantly to environmental degradation. Leather processing involves the transformation of raw materials into finished leather used for various purposes such as footwear, textiles, bags, and more, necessitating a considerable amount of raw materials and chemical treatments [27]. Animal skins primarily consist of water, proteins, and fats, and the leather manufacturing process involves stages like preparatory, tanning, and crusting. During tanning, unstable raw materials are transformed into leather with enhanced strength properties and resistance to physical and biological degradation [28]. Chromium (Cr(III)) salts are commonly used as tanning agents, added after soaking to raise the hide's pH and improve bacterial and thermal resistance. Cr(III) tanned leather typically comprises 2–3% of chromium salts by dry weight. Wet processing contains retanning with chromium, dye addition, and fat inclusion to enhance smoothness and color. After drying and retanning processes, the leather is obtained in the crust form, which then undergoes finishing to achieve softer and cleaner leather products. The finished product contains approximately 65–80% of the dry mass of substances used. Water is predominantly utilized in the tanning procedure within leather industries, with additional water usage in vacuum dryers and throughout various stages of production [29].

1.5.2.5 Pharmaceutical Industry

The pharmaceutical industry significantly contributes to water pollution, utilizing approximately 99% of water for excipient production. The wastewater discharged from drug manufacturing areas contains a range of toxic elements detrimental to both humans and animals [30]. To mitigate these risks, various treatment methods such as membrane filtration, reverse osmosis, flocculation, and coagulation are employed to decrease the presence of elements like BOD, COD, TSS, and TDS in the effluent [31].

1.5.2.6 Agricultural Industry

The commercial agriculture sector consumes significant amounts of water from diverse sources and produces substantial volumes of wastewater. These wastewaters typically contain a range of contaminants including organic matter, dissolved minerals, nutrients like nitrogen, phosphorus, and potassium, toxic chemicals, and pathogens [32]. Various filtration processes are utilized to remove substances such as suspended solids, BOD, COD, bacteria, viruses, fertilizers, manure, and pesticides.

Wastewater treatment in the agriculture industry employs mechanical processes such as filtration, sedimentation, separation, flotation, and crystallization as well as biological techniques such as activated sludge under anaerobic environments. Physicochemical techniques including coagulation, flocculation, electrocoagulation, and ozonation are also employed [33].

1.5.2.7 Textile Industry

The textile industry, one of the largest and most established globally, relies heavily on chemical processing, accounting for approximately 70% of its pollution output. Effluent treatment in this industry focuses on removing suspended solids and improving effluent quality through methods such as screening, sedimentation, neutralization, mechanical flocculation, and chemical coagulation [34]. Secondary treatment targets colloidal and dissolved organic compounds and dye/color removal from wastewater, often employing techniques like aerated lagoons, anaerobic digestion, and trickling filtration. Tertiary treatment further enhances the quality of the slightly contaminated water from previous stages before releasing it into water bodies. This stage involves oxidation techniques, electrolytic precipitation, membrane technologies, and other methods to ensure minimal impact on water bodies [35].

1.5.2.8 Paper and Pulp Industry

The paper and pulp industries play a significant role in our daily lives, producing substantial amounts of biomass [36]. The production processes include debarking, chipping, pulping, and bleaching. Effluent treatment in these industries typically begins with primary techniques such as sedimentation, flotation, and filtration. Secondary biological treatment includes activated sludge processes, aerated lagoons, and anaerobic treatments such as hydrolysis, acidogenesis, acetogenesis, and methanogenesis, as well as sequential treatment methods. Tertiary treatment encompasses membrane filtration techniques like microfiltration, ultrafiltration, nanofiltration, and reverse osmosis, along with coagulation/precipitation and ozonation processes [37].

1.6 RESIDUES AND BIOSOLIDS DISPOSAL

1.6.1 Residues

Residues refer to the solid materials left behind after a process or treatment. They can originate from various sources, including industrial processes, agriculture, and wastewater treatment.

Proper disposal of residues is essential to prevent environmental contamination and health risks. Common methods of disposal include the following:

- **Landfill:** Some non-hazardous residues are disposed of in landfills. However, this method should be used judiciously to avoid overburdening landfills.
- **Incineration:** Incinerating residues can reduce their volume and destroy harmful substances. However, it requires careful management to prevent air pollution.

- **Beneficial reuse:** Some residues, such as ash from incineration or by-products from industrial processes, can be reused in construction materials or other applications [38].

1.6.2 Biosolids

Biosolids are organic materials derived from sewage sludge, which is the solid residue from wastewater treatment plants. Proper biosolids management is crucial for environmental protection and resource recovery. Methods of biosolids disposal and utilization include the following:

- **Land application:** Treated biosolids are applied to agricultural land as fertilizers. This practice enriches the soil and promotes plant growth.
- **Composting:** Biosolids can be composted to create nutrient-rich soil amendments.
- **Energy recovery:** Some facilities convert biosolids into biogas or use them for energy generation.
- **Landfill:** In some cases, biosolids are disposed of in landfills, but this is less sustainable than beneficial reuse.

The proper disposal and utilization of residues and biosolids contribute to a cleaner and healthier environment [39].

1.7 CONCLUSION

The management of water and wastewater is a critical aspect of modern society, touching on environmental sustainability, public health, and economic viability. This overview delved into the fundamental processes and technologies involved in treating water for various purposes, from potable consumption to industrial use. We explored the stages of water treatment, starting from preliminary processes like screening and sedimentation, progressing through coagulation and filtration, and culminating in disinfection to ensure the water is safe for consumption. Wastewater treatment removes pollutants and contaminants from used water before it is returned to the environment or reused. This involves physical, chemical, and biological processes such as primary and secondary treatment, biological nutrient removal, and advanced treatment methods like membrane filtration and oxidation. The goal is to protect human health and the environment as well as to conserve water resources and minimize pollution. Throughout this overview, we also discussed the importance of sustainable practices in water and wastewater treatment. This includes the adoption of energy-efficient technologies, the use of renewable resources, and the implementation of circular economy principles to reduce waste and promote resource recovery. Sustainability also encompasses social and economic aspects, such as ensuring access to clean water for all communities, promoting equitable water distribution, and supporting local economies through water infrastructure development and innovation. Furthermore, we touched upon emerging trends and challenges in water and wastewater treatment, such as the impact of climate change on water availability and quality, the rise of decentralized and nature-based solutions, and the need for resilient infrastructure to withstand extreme weather events. These trends highlight the ongoing evolution and adaptation of water management practices to meet the growing demands and complexities of the modern world. Water and wastewater treatment face challenges such as increasing water demand, ageing infrastructure, and emerging contaminants like pharmaceuticals and microplastics. Sustainable solutions are being pursued, including water reuse and resource recovery from wastewater, to address these challenges and promote circular water management.

REFERENCES

[1] H. Liu, X. Zhao, J. Qu, Electrocoagulation in water treatment, in: C. Comninellis, G. Chen (Eds.), *Electrochemistry for the Environment*, Springer New York, New York, NY, 2010: pp. 245–262. https://doi.org/10.1007/978-0-387-68318-8_10.

[2] O.M. Rodriguez-Narvaez, J.M. Peralta-Hernandez, A. Goonetilleke, E.R. Bandala, Treatment technologies for emerging contaminants in water: A review, *Chem. Eng. J.* 323 (2017) 361–380. https://doi.org/10.1016/j.cej.2017.04.106.

[3] P.P. Das, M. Sharma, M.K. Purkait, Recent progress on electrocoagulation process for wastewater treatment: A review, *Sep. Purif. Technol.* 292 (2022) 121058. https://doi.org/10.1016/j.seppur.2022.121058.

[4] H.Y. Shiu, M. Lee, Z.E. Lin, P. Te Chiueh, Dynamic life cycle assessment for water treatment implications, *Sci. Total Environ.* 860 (2023) 160224. https://doi.org/10.1016/j.scitotenv.2022.160224.

[5] M. Seneviratne, Chapter 2—Basic water chemistry, in: M. Seneviratne (Ed.), *A Practical Approach to Water Conservation for Commercial and Industrial Facilities*, Elsevier, Oxford, 2006: pp. 25–45. https://doi.org/10.1016/B978-185617489-3.50005-9.

[6] N. Colombo, F. Salerno, S. Gruber, M. Freppaz, M. Williams, S. Fratianni, M. Giardino, Review: Impacts of permafrost degradation on inorganic chemistry of surface fresh water, *Glob. Planet. Change* 162 (2018) 69–83. https://doi.org/10.1016/j.gloplacha.2017.11.017.

[7] Z. Jia, T. Liu, X. Xia, N. Xia, Effect of particle size and composition of suspended sediment on denitrification in river water, *Sci. Total Environ.* 541 (2016) 934–940. https://doi.org/10.1016/j.scitotenv.2015.10.012.

[8] K.D. Collins, G.W. Neilson, J.E. Enderby, Ions in water: Characterizing the forces that control chemical processes and biological structure, *Biophys. Chem.* 128 (2007) 95–104. https://doi.org/10.1016/j.bpc.2007.03.009.

[9] J. Esquena, Water-in-water (W/W) emulsions, *Curr. Opin. Colloid Interface Sci.* 25 (2016) 109–119. https://doi.org/10.1016/j.cocis.2016.09.010.

[10] M.K. Purkait, P.P. Das, M. Sharma, Chapter 2—Water treatment techniques for the remediation of drinking and industrial wastewater, in: M.K. Purkait, P.P. Das, M. Sharma (Eds.), *Electrocoagulation Based Treatment of Water and Wastewater*, Elsevier, 2024: pp. 19–46. https://doi.org/10.1016/B978-0-443-13892-8.00011-5.

[11] W.-Q. Guo, S.-S. Yang, W.-S. Xiang, X.-J. Wang, N.-Q. Ren, Minimization of excess sludge production by in-situ activated sludge treatment processes—a comprehensive review, *Biotechnol. Adv.* 31 (2013) 1386–1396. https://doi.org/10.1016/j.biotechadv.2013.06.003.

[12] V.M. Bhandari, L.G. Sorokhaibam, V. V Ranade, Industrial wastewater treatment for fertilizer industry—a case study, *Desalin. Water Treat.* 57 (2016) 27934–27944. https://doi.org/10.1080/19443994.2016.1186399.

[13] G. Mininni, C.M. Braguglia, R. Ramadori, M.C. Tomei, An innovative sludge management system based on separation of primary and secondary sludge treatment, *Water Sci. Technol.* 50 (2004) 145–153. https://doi.org/10.2166/wst.2004.0557.

[14] N.K. Shammas, I.J. Kumar, S.-Y. Chang, Y.-T. Hung, Sedimentation, in: L.K. Wang, Y.-T. Hung, N.K. Shammas (Eds.), *Physicochemical Treatment Processes*, Humana Press, Totowa, NJ, 2005: pp. 379–429. https://doi.org/10.1385/1-59259-820-x:379.
[15] M.R. Rojas, C. Leung, F. Bonk, Y. Zhu, L. Edwards, R.G. Arnold, A.E. Sáez, G. Klečka, Assessment of the effectiveness of secondary wastewater treatment technologies to remove trace chemicals of emerging concern, *Crit. Rev. Environ. Sci. Technol.* 43 (2013) 1281–1314. https://doi.org/10.1080/10643389.2011.644221.
[16] D.P. Zagklis, G. Bampos, Tertiary wastewater treatment technologies: A review of technical, economic, and life cycle aspects, *Processes* 10 (2022). https://doi.org/10.3390/pr10112304.
[17] Y. Zhang, Y. Zhuang, J. Geng, H. Ren, Y. Zhang, L. Ding, K. Xu, Inactivation of antibiotic resistance genes in municipal wastewater effluent by chlorination and sequential UV/chlorination disinfection, *Sci. Total Environ.* 512–513 (2015) 125–132. https://doi.org/10.1016/j.scitotenv.2015.01.028.
[18] S.K. Dubey, A.K. Kashyap, Chapter 14—Biofilm: A doable microbial continuum for the treatment of wastewater, in: G.H. Dar, R.A. Bhat, H. Qadri, K.R. Hakeem (Eds.), *Microbial Consortium and Biotransformation for Pollution Decontamination*, Elsevier, 2022: pp. 321–345. https://doi.org/10.1016/B978-0-323-91893-0.00013-4.
[19] F.J. Trevino Quiroga, Waste stabilization ponds for waste water treatment, anaerobic pond, *Des. Man. Water Stab. Ponds Mediterr.* (2011) 1–23. http://home.eng.iastate.edu/~tge/ce421-521/Fernando J. Trevino Quiroga.pdf.
[20] Y.J. Chan, M.F. Chong, C.L. Law, D.G. Hassell, A review on anaerobic—aerobic treatment of industrial and municipal wastewater, *Chem. Eng. J.* 155 (2009) 1–18. https://doi.org/10.1016/j.cej.2009.06.041.
[21] J.A. Cotruvo, Drinking water standards and risk assessment, *Regul. Toxicol. Pharmacol.* 8 (1988) 288–299. https://doi.org/10.1016/0273-2300(88)90016-5.
[22] P.P. Das, A.D. Sontakke, N.S. Samanta, M.K. Purkait, Emerging contaminants in wastewater: Eco-toxicity and sustainability assessment, in: M.P. Shah (Ed.), *Industrial Wastewater Reuse: Applications, Prospects and Challenges*, Springer Nature Singapore, Singapore, 2023: pp. 63–87. https://doi.org/10.1007/978-981-99-2489-9_4.
[23] M. Changmai, P. Mondal, A. Sinha, P. Biswas, S. Sarkar, M.K. Purkait, Metal removal efficiency of novel LD-slag-incorporated ceramic membrane from steel plant wastewater, *Int. J. Environ. Anal. Chem.* 102 (2022) 1078–1094. https://doi.org/10.1080/03067319.2020.1734198.
[24] N. Shekhar Samanta, P.P. Das, S. Dhara, M.K. Purkait, An overview of precious metal recovery from steel industry slag: Recovery strategy and utilization, *Ind. Eng. Chem. Res.* 62 (2023) 9006–9031. https://doi.org/10.1021/acs.iecr.3c00604.
[25] N.A. Noukeu, R.J. Priso, S.D. Dibong, D. Ndongo, L. Kono, D. Essono, Floristic diversity of receiving environments polluted by effluent from agri-food industries, *Heliyon* 5 (2019) e02747. https://doi.org/10.1016/j.heliyon.2019.e02747.
[26] B. V Raghunath, A. Punnagaiarasi, G. Rajarajan, A. Irshad, A. Elango, G. Mahesh kumar, Impact of dairy effluent on environment—a review, in: M. Prashanthi, R. Sundaram (Eds.), *Integrated Waste Management in India: Status and Future Prospects for Environmental Sustainability*, Springer International Publishing, Cham, 2016: pp. 239–249. https://doi.org/10.1007/978-3-319-27228-3_22.
[27] M.K. Purkait, P.K. Bhattacharya, S. De, Membrane filtration of leather plant effluent: Flux decline mechanism, *J. Memb. Sci.* 258 (2005) 85–96. https://doi.org/10.1016/j.memsci.2005.02.029.
[28] K. Sathya, K. Nagarajan, G. Carlin Geor Malar, S. Rajalakshmi, P. Raja Lakshmi, A comprehensive review on comparison among effluent treatment methods and modern methods of treatment of industrial wastewater effluent from different sources, *Appl. Water Sci.* 12 (2022) 1–27. https://doi.org/10.1007/s13201-022-01594-7.

[29] M. Chowdhury, M.G. Mostafa, T.K. Biswas, A.K. Saha, Treatment of leather industrial effluents by filtration and coagulation processes, *Water Resour. Ind.* 3 (2013) 11–22. https://doi.org/10.1016/j.wri.2013.05.002.
[30] P.P. Das, S. Dhara, M.K. Purkait, Chapter 13—Ozone-based oxidation processes for the removal of pharmaceutical products from wastewater, in: M.P. Shah, P. Ghosh (Eds.), *Development in Wastewater Treatment Research and Processes*, Elsevier, 2024: pp. 287–308. https://doi.org/10.1016/B978-0-443-19207-4.00003-3.
[31] E. Azizl, M. Fazlzadeh, M. Ghayebzadeh, L. Hemati, M. Beikmohammadi, H.R. Ghaffari, H.R. Zakeri, K. Sharafi, Application of advanced oxidation process (H2O2/UV) for removal of organic materials from pharmaceutical industry effluent, *Environ. Prot. Eng.* 43 (2017) 183–191. https://doi.org/10.5277/epe170115.
[32] M.H. Bernier, G.J. Levy, P. Fine, M. Borisover, Organic matter composition in soils irrigated with treated wastewater: FT-IR spectroscopic analysis of bulk soil samples, *Geoderma* 209–210 (2013) 233–240. https://doi.org/10.1016/j.geoderma.2013.06.017.
[33] J.S. Chauhan, S. Kumar, Wastewater ferti-irrigation: An eco-technology for sustainable agriculture, *Sustain. Water Resour. Manag.* 6 (2020) 1–11. https://doi.org/10.1007/s40899-020-00389-5.
[34] P.A. Desai, V.S. Kore, Performance evaluation of effluent treatment plant for textile industry in Kolhapur of Maharashtra, *Univers. J. Environ. Res. Technol.* 1 (2011) 560–565.
[35] M. Bharti, P.P. Das, M.K. Purkait, A review on the treatment of water and wastewater by electrocoagulation process: Advances and emerging applications, *J. Environ. Chem. Eng.* 11 (2023) 111558. https://doi.org/10.1016/j.jece.2023.111558.
[36] M.A. Hubbe, Biomass & energy in pulp & paper, *BioResources* 16 (2021) 8567–8639.
[37] P.P. Das, S. Dhara, M.K. Purkait, Hybrid electrocoagulation and ozonation techniques for industrial wastewater treatment, in: M.P. Shah (Ed.), *Sustainable Industrial Wastewater Treatment and Pollution Control*, Springer Nature Singapore, Singapore, 2023: pp. 107–128. https://doi.org/10.1007/978-981-99-2560-5_6.
[38] M. Taylor, H.A. Elliott, Influence of water treatment residuals on dewaterability of wastewater biosolids, *Water Sci. Technol.* 67 (2013) 180–186. https://doi.org/10.2166/wst.2012.553.
[39] H. Wang, S.L. Brown, G.N. Magesan, A.H. Slade, M. Quintern, P.W. Clinton, T.W. Payn, Technological options for the management of biosolids, *Environ. Sci. Pollut. Res.* 15 (2008) 308–317. https://doi.org/10.1007/s11356-008-0012-5.

2 Introduction to Electrochemistry

2.1 INTRODUCTION

Electrochemistry study focuses on the interface between an electrode and a solution. But most of the basic concepts and relationships were known before J. J. Thompson's discovery of the electron in 1893. In 1800, Alessandro Volta created the first battery, also known as the voltaic pile, by stacking copper and zinc discs in succession and dividing them with paper that had been soaked in acid. The discovery of a workable source of electric current accelerated the development of the scientific subject now known as electrochemistry. By the year 1835, Michael Faraday had established the terms anode, cathode, electrode, electrolyte, and ion; a complete explanation of electrochemistry would be nearly hard to come up with without these [1]. Franklin thought that everything possessed a single, unseen fluid that gave everything its electrical charge; positively charged things had more of this fluid than negatively charged ones. Franklin renamed vitreous and resinous charges as positive and negative, thereby establishing the standard that would later require the electron to be defined as a negative charge. It was Faraday who initially invented the terms ion, anode, and cathode. Conventionally, the anode experiences oxidation, while the cathode experiences reduction [2]. There are two types of cells: galvanic cells, which require batteries, and electrolytic cells, which use applied potential. The two opposing criteria also control the graphical representation of combined voltage/current plots, which are commonly referred to as voltammograms [3].

There are two types of electrochemical cells: galvanic and electrolytic. When the electrodes are connected with a conductor, galvanic cells show impulsive reactions ($\Delta G < 0$) in contrast to electrolytic cells, which require a higher voltage to operate the electrochemical approach. In contrast, the electrodes of an electrolytic cell must fulfill more specific functions than those of a galvanic cell, which only need the anode and cathode to be defined [4]. Counter electrodes (CEs), reference electrodes (REs), and working electrodes (WEs), sometimes known as auxiliary electrodes, are the designations given to the electrodes to help in their differentiation from one another. The WE, which is typically a cathode, is where the reaction of interest typically occurs. Typical examples of WEs include carbon, noble metals like platinum and gold, liquid metals like mercury and the related composites, and others [5, 6]. Chemical changes can be made to WEs to increase their sensitivity to specific species or decrease the potential energy required to execute a given reaction. Regardless of whether an electrolytic cell has two electrodes or three electrodes, it always needs a reference electrode, also known as an anode. A stable potential can be used to observe the WE potential since the RE is composed of a phase arrangement that keeps its configuration nearly constant. The REs are

DOI: 10.1201/9781003499671-2

susceptible to high current densities but enable them to swiftly respond to fluctuations in the ionic activity of the solution. Noble metal electrodes can occasionally be used in place of the reference electrode in cases where there are no appreciable changes in ionic activity [7].

The foundations of electrochemistry, such as reference electrodes, electrochemical cells, and variables influencing reaction rate, are the primary focus of this chapter. A detailed analysis has also been done on the solvent and supporting electrolyte, namely the salts, solvents, and electrolyte conductivity. A thorough discussion of many kinds of electrode materials is also provided in this chapter, including chemically modified electrodes, ultra microelectrodes, semiconductor electrodes, and ion-selective electrodes.

2.2 FUNDAMENTALS OF ELECTROCHEMISTRY

2.2.1 Electrochemical Cells

Electrochemical cells can be made from a wide range of materials. The most often used ones are nylon, Teflon, Kel-F, and glass (quartz and Pyrex). When choosing these materials, it is practical to take into account their cost-effectiveness, ease of processing, and inertness to electrochemical reactions. A Pyrex glass cell is utilized for the majority of electrochemical reactions due to its ease of construction and relatively low cost of glass. The plastic materials may break down in the presence of certain organic solvents, causing contamination and consequent interference with sensitive readings. Glass should always be used instead of nylon when strong acids are present because nylon may not always be stable in such situations. Ultra-small electrodes will need to be employed and correctly aligned in these situations [8].

Nanoparticles and other materials' solid-state electronic conductivity has also been measured using two-electrode cells. A typical cell is a concentric tubular construction with an inner diameter of a few millimeters. It contains an electrolyte that is contained within a tubular interior and is sandwiched between two disk-shaped electrodes that can be plugged in. The materials used in the cell should therefore be able to withstand a fair amount of mechanical pressure. When the cell resistance is rather large, the most popular design for electrochemical investigations is three-electrode cells. In this setup, current flows between the working electrode and a different auxiliary (counter) electrode, but the working electrode's potential is still monitored in relation to the reference potential [9, 10]. The reference electrode approaches optimum non-polarizability since no (or very little) current flows to it; as a result, it serves as a trustworthy reference for potential control. Control of the voltammetric current scale will also be crucial, and in order to minimize (uncompensated) solution resistance, the reference electrode tip is positioned as close to the working electrode as feasible in experiments. A medium-porosity glass frit can be used to divide the three compartments that typically make up a cell. The supporting electrolyte and redox molecules are typically introduced into the center compartment, while the supporting electrolyte is simply added to the two side compartments. To facilitate degassing, an additional aperture might be installed [11].

2.2.2 Reference Electrodes

The selected RE should be a perfect non-polarizable electrode, meaning that no matter how much current passes through it, its potential will remain constant. To determine the potential using the Nernst equation, the reaction at the RE must be reversible. It should also be able to regain its potential following a current stress [12]. For the length of the experiment, the RE must function as a source or a sink for tiny currents caused by flaws in the potentiostat or galvanostat design without showing a change in potential. When selecting a good RE for an experiment, several factors need to be taken into account. The equilibrium of the redox pair depends on various conditions in the cell, including temperature and pressure. If the RE remains constant during the measurement, time constraints may permit the adoption of a less robust RE. However, there are a number of RE types that can be used in electrochemical tests [13, 14].

Non-aqueous solvent systems also bring new factors into the equation when choosing a RE for electrochemical experiments. Aqueous REs may be utilized under specific conditions, but the electrolyte must dissolve in a non-aqueous solvent. Using a salt bridge or double-junction RE is an additional method of employing an aqueous RE in a non-aqueous solution. Tetraethyl ammonium picrate and tetra-n-butyl ammonium tetraphenylborate can both be utilized in salt bridges since they are almost equitransferent in a variety of organic solvents. The potential measurements that are made, even when employing the same RE, cannot be related from one solvent to another when using aqueous REs in non-aqueous solvents [15]. This is because different systems have different junction potentials. Numerous organic solvents can be utilized with the Ag|Ag+ RE. The junction potential of the RE will be reduced if the same solvent is used in the electrochemical cell and the RE filling solution. This RE has the same drawback as that faced when utilizing aqueous REs to compare potentials between various solvents. One method of lowering the uncompensated resistance is to sample the electrochemical cell near the WE using a Luggin capillary, making sure the capillary's output is close to the WE. The Luggin capillary and the WE should be at least twice as far apart as the WE's diameter in order to avoid shielding the WE [16].

2.2.3 Cell Resistance and Capacitance

Both faradaic and non-faradaic currents can flow in electrochemical cells. Any current generated by the oxidation or reduction of chemical complexes inside the chamber is referred to as a faradaic current. It is possible to model faradic currents using the below equation [17]:

$$\frac{dQ}{dt} = \frac{d}{dt}(\mathrm{nFN}) = \mathrm{nF}\frac{dN}{dt}$$

Here, Q stands for charge (C), n for the number of transferred electrons, t for time (s), N for the number of reacting moles, and F for Faraday's constant.

In order to comprehend the effects of electrode capacitance and resistance in an electrochemical cell, it is helpful to construct the concept of ideal polarizable electrodes (IPE). The IPE's behavior then resembles that of a capacitor; however, in contrast to a conventional capacitor, the IPE's capacitance is dependent upon the applied potential across it. An electrical double layer is produced by the cell's non-faradaic current supply and the application of voltage across the electrodes. The electrode surface is X_1 away from the inner Helmholtz plane (IHP) boundary, where this layer's electrical centers are situated [11]. The uppermost layer is composed of solvated ions. The electrical charges from these ions are located in the outer Helmholtz plane (OHP), as they are completely encircled by solvent molecules and can only approach X_2. Ions that have not been particularly adsorbed are still solvated and can only approach the OHP. Only at long range do the non-specifically entrapped ions interact with the electrode surface; their chemical makeup has no bearing on these interactions. Less compact than the IHP, the ionic species in this plane are distributed throughout the diffuse layer, a three-dimensional region that is firmly ingrained in the bulk solution [18].

2.3 SOLVENT AND THE SUPPORTING ELECTROLYTE

2.3.1 Solvents Used in Electrochemistry

In addition to water, a wide variety of organic liquids have been employed as solvents in electrochemical studies. A wealth of comprehensive literature has been written about the characteristics of solvents and solvent–electrolyte combinations that can be utilized as media in electrochemical experiments. The most crucial solvent qualities might vary greatly and are determined by the application. Nevertheless, a few categories of solvent properties that are particularly significant for electrochemical applications can be the focus of attention. It is evident that there exist issues related to the utilization of solvents lacking sufficient ionic conductivity [19]. Actually, a solvent's solvating power—a more general solvent property—is what makes it possible for it to dissolve and dissociate salt. Solvents are typically used in electrochemical reactions to solvate the reactants and products. This is why solvents are frequently selected for this purpose. Optical qualities, volatility, toxicity, liquid range, and cost are additional variables that may be significant [20]. A list of some of the most often used solvents in electrochemistry is provided in Table 2.1. The structures of some of these solvents are shown in Figure 2.1, where they are arranged to emphasize the structural similarities between and among different solvents.

As acetonitrile is completely miscible with water, it might be helpful in research where a mixed solvent system is preferred due to solubility or reactivity issues. Their (acetonitrile) use has decreased recently because of worries about halogenated solvent disposal and their toxicity, which definitely restricts their use in any large-scale applications. Nevertheless, they remain popular because they offer an excellent balance between polarity and organic solvating power for research applications [21]. Since halogenated solvents are typically susceptible to electroreduction, their use may be restricted when strong reducing agents or highly reducing potentials are used. These are excellent options for electrochemical oxidation. N-methylpyrrolidinone

TABLE 2.1
Characteristics of some frequently used solvents in electrochemistry (Reproduced with permission from Creager (2007) @ Elsevier [8]).

Common name	Molecular weight (g/mol)	Freezing point (°C)	Boiling point (°C)	Density (g/cm³)	Viscosity (centipose)	Dielectric constant (DC)	Dipole moment
Water	18.02	0.0	100.0	0.997	0.890	80.1	1.85
Methanol (MeOH)	32.04	-97.5	64.6	0.791	0.544	33.0	1.70
Acetone	58.08	-94.7	56.1	0.785	0.306	21.0	2.88
Acetonitrile (AN)	41.05	-43.8	81.7	0.786	0.369	36.6	3.92
Butyronitrile	69.11	-111.9	117.6	0.794	0.553	24.8	3.9
Benzonitrile	103.12	-14.0	191.1	1.009	1.267	25.9	4.18
Dichloromethane	84.93	-97.2	40.0	1.327	0.413	8.93	1.6
Tetrahydrofuran (THF)	72.11	-108.44	65	0.883	0.456	7.52	1.75
Dimethylsulfoxide (DMSO)	78.13	17.89	189	1.101	1.987	47.2	3.96
N,N-dimethyl formamide (DMF)	73.09	-60.5	153	0.945	0.794	38.3	3.82

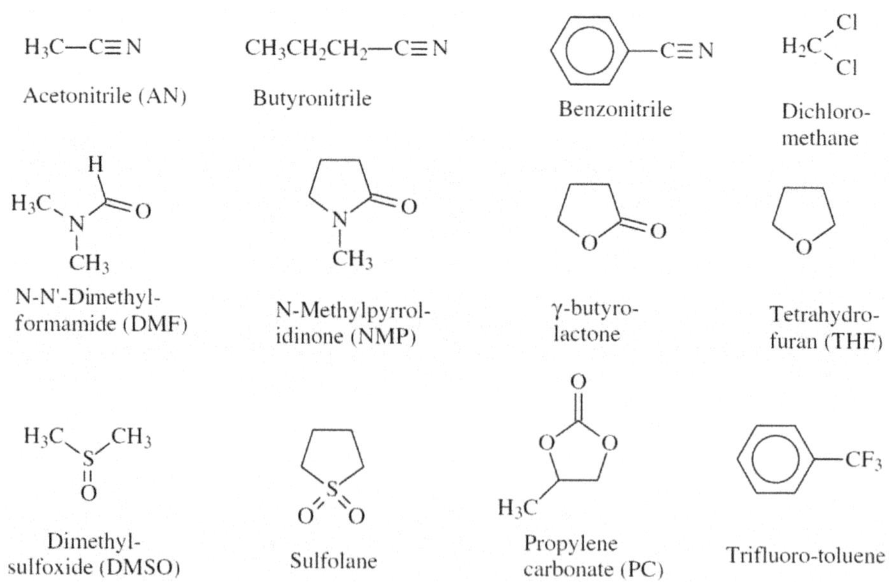

FIGURE 2.1 Structures of organic solvents commonly used in electrochemistry (Reproduced with permission from Creager (2007) @ Elsevier [8]).

(NMP) is a multiused polar organic solvent and the cyclic analog of DMF. Sulfolane and dimethyl sulfoxide (DMSO) are also suitable options if a polar, high-boiling solvent is required. Although DMSO is highly polar and readily dissolves a wide range of salts and organic molecules, its high viscosity restricts transport rates and ionic conductivities. Sulfolane is a valuable organic solvent for high-temperature electrochemical research since it has one of the highest boiling points among all organic solvents [22].

2.3.2 Salts in Electrochemistry

Considering the total number of salts known, it is surprising how few are commonly used as supporting electrolytes in electrochemical experiments. This is especially true for organic solvents, which are generally not thought of as good solvents for salts and for which salts typically require substantial organic characteristics to be soluble [23]. In these types of salts, the anions and cations can be thought of as separate entities to some extent because both species are generally stable and can typically be "mixed and matched" to produce the desired salt as shown in Figure 2.2. Tetra-alkyl ammonium ions are by far the most widely employed cations; their organic nature along with their comparatively low electro-reductive reactivity makes them ideal for use in electrochemical investigations. The most commonly used ions are tetraethyl ammonium (TEA) and tetrabutyl ammonium (TBA). The anions are more varied than the other groups. Although perchlorates were once

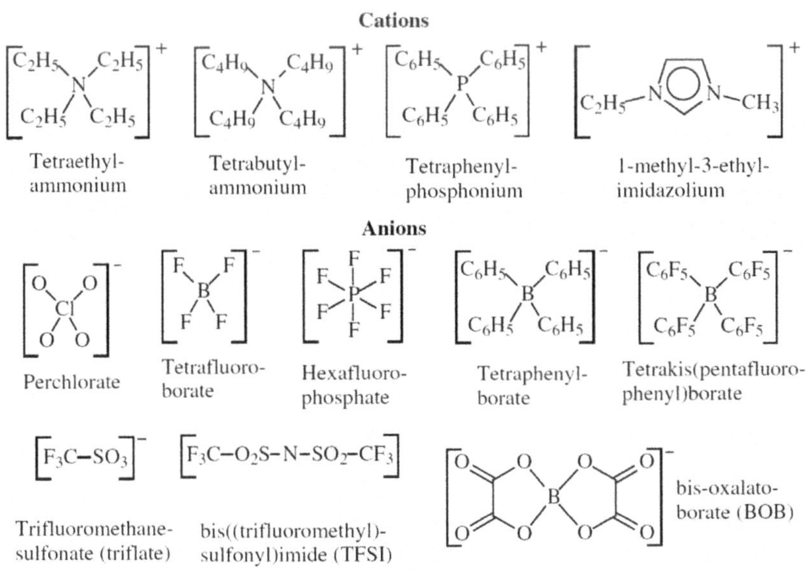

FIGURE 2.2 Frequently used anion and cation salts in electrochemistry (Reproduced with permission from Creager (2007) @ Elsevier [8]).

widely employed, their use is now discouraged because of potential explosion hazards. Salts of tetra fluoroborate and hexafluorophosphate are commonly utilized as good substitutes for perchlorates [24].

2.3.3 Electrolyte Conductivity

Supporting current flow is one of the most crucial characteristics a medium need to have in order to be effective in electrochemical science. Ions are always produced or consumed at electrodes during electrochemical reactions, and the electrolyte creates the passageway for ions to move between and among electrodes within the cell in order to preserve charge balance. The term conductivity, sometimes denoted by κ, refers to a phase's capacity to sustain current flow caused by the migration of charged particles (ions in an electrolyte) when an electric field is present inside the phase. The unit of conductivity is ohm^{-1} cm^{-1}, often known as siemens cm^{-1}. A phase or material's conductivity can be determined using the formula $\kappa = (1/R)(L/A)$ for an electrolyte phase with a uniform cross section of area A and length L and an electrical resistance R [25]. In terms of ohm cm, conductivity is the inverse of resistivity and is commonly represented by the symbol ρ. It is outside the purview of this chapter to address every aspect of ionic conductivity in electrolytes. Nevertheless, it is useful to have a cursory look at the variables influencing ionic conductivity in order to comprehend how the properties of the solvent and electrolyte affect the conductivity of different solvent–electrolyte combinations. Very concentrated salt solutions may have a high viscosity, which may restrict ion mobility and result in reduced conductivity. Strong and weak electrolytes are the two general categories. Strong electrolytes are typically used when combining solvent and electrolyte for use in electrochemical studies. Ion pairing can be thought of as a conflict between the heat energy in the medium, which acts to separate the ions, and the energy of electrostatic attraction of two oppositely charged ions toward each other, which acts to keep the ions together [8]. Nevertheless, it is helpful to think about certain theoretical expectations about how the characteristics of the solvent and electrolyte should affect ion pairing. The earliest and most basic theory of ion pairing, known as the Bjerrum theory, states that ion pairing will be highest under conditions of high concentration of salt, small and highly charged salt ions, low temperature, and low solvent permittivity (dielectric constant). An intriguing trade-off is shown when ion size is taken into account in relation to electrolyte conductivity. It would seem that large ions are not preferred because of their poor mobility [26].

2.4 TYPES OF ELECTRODE MATERIALS

2.4.1 Chemically Modified Electrodes

In a chemically modified electrode (CME), the adlayers have a wide range of sources and characteristics, whereas the conductive and semiconductive substrates are made of traditional electrode materials. The breadth and depth of electrochemical procedures are expanded or improved by this diversity. Ideally, the electrode's function is predictable due to the characteristics of the adlayer. Advancements in surface characterization methodologies have made it possible to comprehend changed interfaces at the molecular level. In addition to offering a way to confirm the function, these methods, when combined with electrochemical characterization techniques, also create the foundation for improving the modification strategy's performance. An exhaustive description of all modified electrode types, their applications, and the methods employed to characterize them is not possible due to the vast nature of electrode modification procedures [27]. Thus, the purpose of this chapter is to introduce the reader to key principles of electrode modification using multiple real-world examples of systems. The platform electrode that holds up the adlayer is called the substrate. Substrates that can be purchased commercially or made to order can be used. Although a substrate can be made of any common electrode material, some are easier to work with than others. The most often changed conductive metals are Au, Ag, and Pt. The simplicity with which clean, repeatable surfaces may be created and maintained in a laboratory setting is advantageous to these surfaces [28]. Carbon electrodes are also frequently used for comparison purposes. Even though they are used less frequently, semiconductors and other metals are still well-known materials. To ensure repeatable surface characteristics, the electrode surface is treated prior to modification in the majority of electrode preparation schemes. Conventional electrode treatments are frequently used in these procedures, with modifications or additions made as necessary. Pretreatment is essential because even slight changes in the substrate's characteristics (such as crystallinity, roughness, or chemical functionality) can have a significant impact on the electrochemical performance of a derivatized electrode. For preparatory techniques to operate as intended, they must therefore often be fine-tuned via trial and error. However, by adjusting pretreatment variables, the electrode's performance may typically be optimized, increasing the adaptability of a modification strategy [29].

The film composition (such as clays, zeolites, sol-gels, polymers, and DNA) and the film–substrate attachment mechanism can be used to categorize modified electrodes. The final application's requirements determine the alteration techniques and film identity to be used. It is possible to realize various combinations of film compositions and attaching techniques. For the purpose of preparing a modified electrode, the Langmuir-Blodgett (L-B) assembly process provides a flexible method. On the electrode surface, attractive noncovalent interactions keep the film together as a cohesive unit. However, adlayer stability is limited by the feebleness of these forces. Because L-B films are easily made to resemble lipid membranes, they are useful in biological studies. For this reason, studies of transport processes and enzyme activity frequently employ modified electrodes of this kind [11]. These electrodes are also frequently utilized in molecular electronics and as sensors for molecule recognition. A Langmuir film is deposited onto a solid substrate to generate Langmuir-Blodgett films. An ordered monolayer of amphiphilic molecules at a liquid/gas contact makes up Langmuir films. These molecules are naturally orientated at a liquid/gas interface due to their hydrophobic tail and hydrophilic head. amphiphilic mixtures can be spread upon a common solvent to create films with a mixed composition [30]. The chemical makeup of the molecules dispersed at the interface produces homogeneous or phase-segregated films. To raise the two-dimensional film concentration, lateral pressure is applied once the desired composition—single or multi-component—is attained. In actual use, a movable barrier within a Langmuir trough is used to apply pressure. Phase transitions occur in the system as a result of compression of the film. A very well-organized two-dimensional film is produced at the conclusion of the compression cycle. The compressed film can be placed on an electrode surface after it has been created. A common technique for film transfer involves lifting a hydrophilic substrate vertically through a Langmuir film. The layer transfers as a result of the attractive interactions between the head groups and the surface, causing the film to physisorb on the electrode surface. These pressures are frequently enough to keep solvent molecules away from the solid support-film contact. By extending the deposition process downward from the gas phase into the liquid layer via the Langmuir film, the head groups are oriented away from the surface. Repeated cycles of raising and/or lowering the substrate through the Langmuir film can deposit many layers [1].

2.4.2 Semiconductor Electrodes

Particularly in physical chemistry, electrochemistry using semiconductor electrodes is an intriguing and extremely interdisciplinary field of study. It is instructive to look first at this field's historical development. E. Becquerel made the first demonstration of using a semiconductor as an electrode in an electrochemical cell in 1839. When UV light was applied to the AgCl electrode, he noticed a photovoltaic effect. It was not until 1954 that scientists at Bell Laboratories examined photo electrochemistry on Ge single crystals that the "Becquerel effect" was fully recognized. They showed that the amounts of impurities in Ge have a significant impact on the photochemical reactions that take place at the electrodes. These experiments, along with the then-emerging zone theory, led to the modeling of the Becquerel effect as a photo-induced charge separation at the AgCl/liquid interface. Additional semiconductor electrodes, including Si, GaAs, GaP, CdS, CdSe, ZnS, and ZnSe, among others, were used to examine the effect further [11]. During this time, theories regarding electron transfer were also developing quickly. The work was basic in nature and primarily of scholarly significance until the early 1970s. However, the 1972 study by Fujishima and Honda on photo-assisted H_2O splitting on TiO_2 raised the potential of employing semiconductor electrodes to transform solar energy into chemical or electrical energy [31]. The process of electrolyzing water with sunlight is very appealing because (a) the hydrogen produced may be stored more readily than heat or electricity; and (b) it is a renewable, inexhaustible, non-polluting fuel that is far more versatile than traditional fuel choices. The world was witnessing skyrocketing gas costs for the first time because to the depletion of fossil fuels at the time this report was released. As a result, there is currently a lot of research being done to better understand the semiconductor/electrolyte interface (SEI) in relation to renewable energy sources. The aim is to find the optimal SEI that can provide an efficiency of roughly 15% without damaging the electrodes. The inverse link between photocorrosion and efficiency has limited success to date. Although oxide semiconductors with a large band gap, such as ZnO and TiO_2, are somewhat resistant against photocorrosion, they have low efficiency. Due to their ideal band gaps, selenides and sulfides are more efficient, but they become unstable in the presence of sunlight [32]. Semiconductor electrodes offer an intriguing substitute for conventionally employed working electrode materials in electrochemical research, notwithstanding their limited performance in resolving energy-related issues. The ability to easily alter the electrochemical response of semiconductor electrodes by adjusting impurity levels, surface treatments, or surface adsorption is a significant benefit. Because of this, semiconductor electrodes offer a significant advantage over traditional metal electrodes, where it is more difficult to manage a particular metal electrode's characteristics. This characteristic of semiconductor electrodes may be useful in the creation of sensors. In an electrochemical cell, a measured current is typically understood to represent charge transfer across the electrode—solution contact. As a result, theoretical calculations do not require the usage of individual electrons or holes. The anisotropic crystallographic characteristics of layered metal dichalcogenide (LMD) semiconductors have a significant impact on the bulk and surface characteristics of the electrodes. However, by merely layering the top few layers with the aid of adhesive tapes, the molecularly smooth, chemically inert, and reproducible van der Waals surface of LMDs may be achieved with ease. With the exception of a handful that are frequently utilized by the electronic industries, such as Si, GaAs, and InP, single crystals of semiconductors are frequently difficult to obtain. A film known as an epitaxial layer can occasionally be grown on a substrate with matching lattice characteristics that is a few atomic layers thick. For this, methods like metal organic chemical vapor deposition (MOCVD) and molecular beam epitaxy (MBE) have been used [33]. A semiconductor single-crystal membrane can also be grown with long-range ordering, resulting in a thick sheet as thick as a few micrometers. The high preparation costs associated with single-crystal and epitaxial films represent a significant drawback to their use as electrodes. As a result, polycrystalline electrodes have been used in the majority of research. Electrodes made of numerous semiconducting materials with moderate individual crystallite sizes relative to the charge carrier diffusion length are comparatively easy to produce. The dispersion of these charge carriers along the grain boundaries, or crystallite borders, significantly hinders the use of the Gerischer model for charge transfer [34]. Furthermore, there is a significant reduction in the efficiency of charge transfer, making it challenging to quantify the electrical characteristics of polycrystalline semiconductor electrodes. The production and characterization of polycrystalline electrodes have been well studied in spite of this issue. This interest stems from polycrystalline electrodes with a large surface area that can be manufactured at a reasonable cost. The steps involved in creating these electrodes are covered in the sections that follow. The impact of size quantization on the electronic characteristics of semiconductors indicates the significant practical utility of semiconductor electrodes composed of nanostructured particles. These films can be divided, according to size quantization, into two categories: (a) thin semiconductor films epitaxially grown on a substrate, where the SQE results from space confinement in two dimensions (a quantum well); and (b) particulate films of size-quantized nanoparticles, which may have a thickness of several micrometers; the properties of these films are due to the interaction between the film and isolated size-quantized particles [3].

2.4.3 Ultra Microelectrodes

Electrodes with distinctive sub- or micrometer-scale dimensions are known as ultra microelectrodes, or UMEs. They have revolutionized thermodynamic, kinetic, and electroanalytical measurements with faster double-layer charging, lower ohmic loss, and higher mass-transport rates. Due to these characteristics, electrochemistry is now able to operate in settings that are not achievable with conventional electrodes, such as liquid CO_2 and single living cells, as well as short length scales and nanosecond timescales [35]. The effective use of microelectrodes is dependent on these decreased currents:

> Unusual media: Because electrochemical measurements could only be performed in highly ionic conducting fluids, electrochemistry has historically lagged behind spectroscopy. This limitation resulted from resistance between the reference electrode and the working or sensing electrode, which made it difficult to precisely adjust the applied potential. These ohmic effects are frequently entirely eliminated by the tiny electrolysis currents seen at microelectrodes.

Small volumes: When working with small reaction volumes (such as those found in capillary electrophoresis separation systems and biochips), electrochemical technologies have a number of benefits over spectroscopy. A tiny sample volume does not impair the limits of detection because electron transfer takes place at an electrode surface. In contrast, the signal intensity in absorption spectroscopy is linearly proportional to the path length. From a biological standpoint, it is possible to examine individual cells and track responses in minuscule amounts owing to the efficient execution of voltammetric measurements in picoliter vials [36].

In their examination of cyclic voltammetry, hydrodynamic voltammetry, and sono voltammetry for evaluating the kinetics and processes of electrode reactions, Eklund et al. (1999) touched on the subject of mass transfer modeling [37]. Here, we concentrate on uses that make use of the special qualities of microelectrodes, from electroanalysis in electrolyte-free solutions to measurements in small volumes. The development of small-volume, real-time, and spatially resolved measurement methods in analytical chemistry has been greatly aided by the use of microelectrodes. Numerous fields of electroanalysis, such as environmental, biological, and material research, take advantage of the benefits of microelectrodes. Micro electrochemistry is a fast-growing discipline that connects fundamental physics to microelectronics, corrosion research, medicine, and biology, as shown in Figure 2.3. Microelectrodes' small physical size enables the electrochemical cell's dimensions to be drastically decreased, enabling direct measurements in nanoliter and even picoliter quantities. For micro electrochemical measurements, a variety of electrochemical cells have been created. The cell volume, electrode spacing, electrolyte flow, and optical transparency are important characteristics. Apart from nanolithography, Bard and Fan (1995) devised a method to identify individual redox molecules by guiding a microelectrode toward a sizable counter electrode, so forming a micro volume electrochemical cell [38]. A redox active molecule can be electrolyzed repeatedly to produce a detectable current by being transferred from bulk solution into the micro volume formed by the two electrodes.

The concentration of a variety of analytes, particularly metal ions, has been measured using microelectrodes in conjunction with traditional analytical methods such

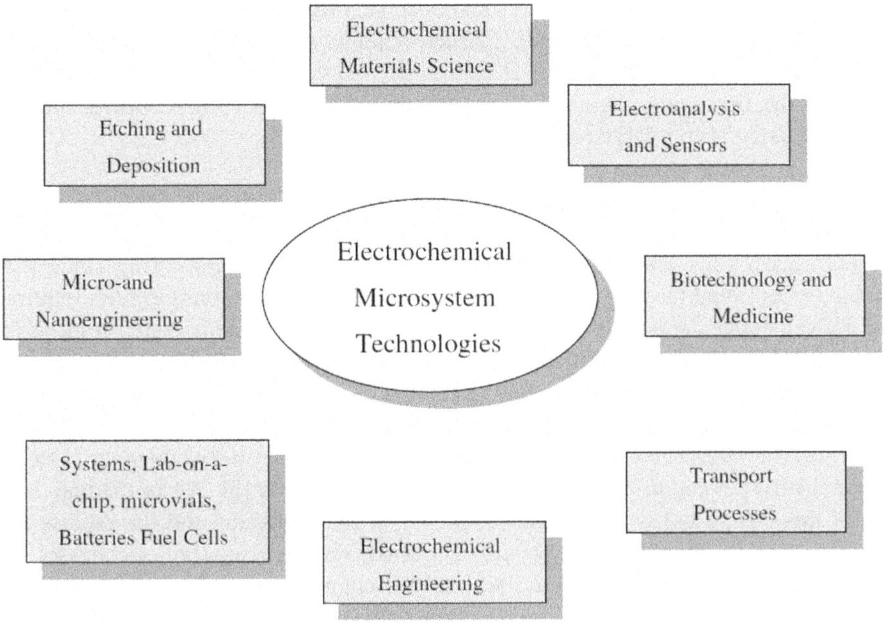

FIGURE 2.3 Schematics of electrochemical microsystem techniques (Reproduced with permission from Forster and Keyes (2007) @ Elsevier [39]).

as anodic stripping voltammetry (ASV). Here, however, we concentrate on more modern methods that create sensors by altering the microelectrode's surface. The structural element of cholesterol governs the fluidity and permeability of the plasma membrane found in mammalian cells. It plays a crucial role in immunological response, cell infection, and cell surface polarity in addition to signal transmission within cells [40]. Devadoss and Burgess (2004) used cholesterol oxidase-incorporated lipid bilayer-modified microelectrodes to investigate the steady-state detection of cholesterol in a single cell's plasma membrane [41]. The steady-state electrode response in their single-cell method appears to be correlated with the amount of cholesterol in the cell plasma membrane. The response is intermediate when the electrode is positioned exactly next to the cell, and maximal current is obtained when the electrode makes some forceful contact with the cell. The reaction that was seen next to the cell might be an indication that cholesterol efflux, or solution-phase cholesterol, has been detected. Initially, a glass capillary's end is sealed to create a conical shape. With the aid of a heated resistor coil, the capillary is gradually sealed onto the wire. Next, a conductive silver epoxy is used to electrically join the sealed wire to a bigger wire. To cure the epoxy, the attached tip is kept in an oven at 120°C for the entire night. Moreover, this process can be used on 5 and 10 μm platinum and gold wires. Before sealing the tip, a mild nitric acid solution is used to dissolve the silver layer on a Wollaston wire, which is a metal wire wrapped in a layer of silver, for tiny tips with a diameter of 1–2 μm. Submicron size electrodes can also be created using a laser puller and tiny quartz capillaries. Such sealing methods call for a great deal of patience and practice. Electrode characterization, polishing, and sharpening are further steps in the production of UME tips. To assess the quality of the glass/metal seal and the tip radius, the electrode's voltammetric behavior is recorded, and the recorded value is compared to the optical observation [42].

2.5 SPECIFIC APPLICATION OF ELECTROCHEMICAL CELLS

Flow-through cells are capable of real-time detection and monitoring since they include a continual feeding of sample solution. The particular configuration required for a given experiment varies. There are two- and three-electrode configurations for flow cells. For use in flow cell applications, a broad range of electrode materials are available. In this setup, a steady-state current is measured while the sample solution is continuously delivered through the cell. Note that a cation-exchange membrane separates the reference electrode from the sample solution in this instance, and the internal filling solution is also changed on a regular basis [3]. An immediate benefit of this design is the significant simplicity of the apparatus and the precise and accurate monitoring of low analyte concentrations. When using three electrodes, the working electrode is positioned in the thin flow channel's walls opposite to the auxiliary electrode, and the flow direction is usually parallel to the working electrode surface. Because of the low impedance between the electrodes, the IR drop has no effect on the current density across the electrode. Using working electrodes in series or parallel can improve performance even more [11].

A micrometer-thick (2–100 μm) layer of electrolyte solution is usually limited at the electrode surface in thin-layer cells (TLCs); the thin-layer flow cell described above is one example. Mass transfer within the layer can be disregarded because the cell thickness is typically less than the diffusion layer thickness. As a result, the process is similar to bulk electrolysis but takes place in a small interfacial area. A micrometer scale allows for precise adjustment and measurement of the solution layer's thickness. The two working electrodes in a twin-electrode (generator–collector) setup should preferably have the same geometric area and be oriented concentrically toward each other. With its electrode surface level aligned with the bottom of the compartment, the bottom electrode is implanted in the cell housing [1]. A set screw, for example, can be used to secure the entire housing to the stationary end of a micrometer scale. The micrometer scale's moving piston is connected to the top electrode. To serve as an enclosure, the cell chamber may have a cover placed on it. A hole for the top working electrode is drilled in the center, and two tiny holes are made for the reference and counter electrodes' electrical connections (a second hole may be used for deaeration purposes). Placing the electrode surface level and ideally parallel to each other is crucial [18].

2.6 CONCLUSION

The multidisciplinary area of electrochemistry, which includes physics, chemistry, and materials science, depends on a wide range of elements and ideas in order to explain the behavior of electrochemical cells and develop different applications. We have covered a wide range of topics in this thorough investigation, from basic theories to cutting-edge techniques. The foundation of electrochemical research are electrochemical cells, which provide information about redox reactions and energy conversion. For applications ranging from energy storage to corrosion prevention, their design, optimization, and comprehension are essential. Additionally, reference electrodes are essential for maintaining a steady reference potential that allows for precise measurements, guaranteeing the dependability and repeatability of electrochemical investigations. It becomes clear that cell resistance and capacitance are important factors affecting how well electrochemical systems function. Researchers can make significant advancements in fields like sensor technology and electroanalytical chemistry by improving the sensitivity and reliability of electrochemical measurements through an understanding of and mitigation of these effects. Salts and solvents have a big impact on electrochemistry because they affect reaction kinetics and thermodynamics in addition to acting as media for electrolyte solutions. Comprehending their characteristics and interplay is essential for customizing electrolytes for particular uses and enhancing electrode procedures. One key parameter determining the effectiveness of ion transport in electrochemical cells is electrolyte conductivity. Researchers continuously push the limits of conductivity through improvements in electrolyte engineering and design, enabling high-performance electrochemical devices with improved functionality and durability. An adaptable platform for customizing surface characteristics and improving electrochemical performance is provided by chemically modified electrodes. Researchers may create electrodes with specific functions by utilizing surface chemistry and nanoengineering techniques, opening up possibilities for everything from energy conversion to sensing. At the nexus of solid-state physics and electrochemistry, semiconductor electrodes present fascinating prospects. Using the distinct electrical characteristics of semiconductors, scientists investigate new approaches for photo-electrochemical energy conversion and electrocatalysis, propelling advancements in the direction of sustainable energy solutions. Among the most effective instruments for examining electrochemical reactions at the nanoscale are ultra microelectrodes. Their compact size provides previously unheard-of levels of spatial resolution and sensitivity, paving the way for novel applications in micro electrochemistry and biosensing and deeper understanding of basic electrochemical processes.

To sum up, the intricacies involved in the interaction of electrochemical cells, semiconductor electrodes, ultra microelectrodes, solvents, salts, cell resistance and capacitance, and chemically modified electrodes with electrolyte conductivity highlight the depth and intricacy of electrochemistry. Researchers open the door for revolutionary discoveries in a variety of domains, such as energy storage and conversion and biosensing, by deepening our grasp of these elements and ideas.

REFERENCES

[1] C.G. Zoski (Ed.), *Handbook of Electrochemistry*, Elsevier, Amsterdam, 2007. ISBN: 978-0-444-51958-0.

[2] P.P. Das, S. Dhara, M.K. Purkait, Hybrid electrocoagulation and ozonation techniques for industrial wastewater treatment, in: M.P. Shah (Ed.), *Sustainable Industrial Wastewater Treatment and Pollution Control*, Springer Nature Singapore, Singapore, 2023: pp. 107–128. https://doi.org/10.1007/978-981-99-2560-5_6.

[3] S. Chen, Practical electrochemical cells, in: C.G. Zoski (Ed.), *Handbook of Electrochemistry*, Elsevier, Amsterdam, 2007: pp. 33–56. https://doi.org/10.1016/B978-044451958-0.50003-3.

[4] P.P. Das, A. Anweshan, M.K. Purkait, Treatment of cold rolling mill (CRM) effluent of steel industry, *Sep. Purif. Technol.* 274 (2021) 119083. https://doi.org/10.1016/j.seppur.2021.119083.

[5] J. Giner, A practical reference electrode, *J. Electrochem. Soc.* 111 (1964) 376–385. https://doi.org/10.1149/1.2426125.

[6] P.P. Das, S. Dhara, M.K. Purkait, Ozone-based oxidation processes for the removal of pharmaceutical products from wastewater, in: M.P. Shah, P. Ghosh (Eds.), *Development in Wastewater Treatment Research and Processes*, Elsevier, 2024: pp. 287–308. https://doi.org/10.1016/B978-0-443-19207-4.00003-3.

[7] I.B. Goldberg, A.J. Bard, Simultaneous electrochemical-electron spin resonance measurements. Part I: Cell design and preliminary results, *J. Phys. Chem.* 75 (1971) 3281–3289. https://doi.org/10.1011/J.phyche-04608.0014.

[8] S. Creager, Solvents and supporting electrolytes, in: C.G. Zoski (Ed.), *Handbook of Electrochemistry*, Elsevier, Amsterdam, 2007: pp. 57–72. https://doi.org/10.1016/B978-044451958-0.50004-5.

[9] K.R. Kneten, R.L. McCreery, Effects of Redox system structure on electron-transfer kinetics at ordered graphite and glassy carbon electrodes, *Anal. Chem.* 64 (1992) 2518–2524. https://doi.org/10.1021/ac00045a011.

[10] P.P. Das, P. Mondal, Anweshan, A. Sinha, P. Biswas, S. Sarkar, M.K. Purkait, Treatment of steel plant generated biological oxidation treated (BOT) wastewater by hybrid process, *Sep. Purif. Technol.* 258 (2021). https://doi.org/10.1016/j.seppur.2020.118013.

[11] G.M. Swain, Solid electrode materials: Pretreatment and activation, in: C.G. Zoski (Ed.), *Handbook of Electrochemistry*, Elsevier, Amsterdam, 2007: pp. 111–153. https://doi.org/10.1016/B978-044451958-0.50006-9.

[12] P.P. Das, A. Anweshan, P. Mondal, A. Sinha, P. Biswas, S. Sarkar, M.K. Purkait, Integrated ozonation assisted electrocoagulation process for the removal of cyanide from steel industry wastewater, *Chemosphere* 263 (2021) 128370. https://doi.org/10.1016/j.chemosphere.2020.128370.

[13] P.P. Das, A.D. Sontakke, M.K. Purkait, Electrocoagulation process for wastewater treatment: Applications, challenges, and prospects, in: M.P. Shah (Ed.), *Development in Wastewater Treatment Research and Processes*, Elsevier, 2023: pp. 23–48. https://doi.org/10.1016/B978-0-323-95684-0.00015-4.

[14] P.P. Das, M. Sharma, M.K. Purkait, Recent progress on electrocoagulation process for wastewater treatment: A review, *Sep. Purif. Technol.* 292 (2022) 121058. https://doi.org/10.1016/j.seppur.2022.121058.

[15] P. Chen, R.L. McCreery, Control of electron transfer kinetics at glassy carbon electrodes by specific surface modification, *Anal. Chem.* 68 (1996) 3958–3965. https://doi.org/10.1021/ac960492r.

[16] M. Josowicz, J. Janata, M. Levy, Electrochemical pretreatment of thin film platinum electrodes, *J. Electrochem. Soc.* 135 (1988) 112. https://doi.org/10.1149/1.2095533.

[17] M. Ciobanu, J.P. Wilburn, M.L. Krim, D.E. Cliffel, Fundamentals, in: C. Zoski (Ed.), *Handbook of Electrochemistry*, Elsevier, 2007: pp. 3–29. https://doi.org/10.1016/B978-044451958-0.50002-1.

[18] P. Sun, M.V. Mirkin, Kinetics of electron-transfer reactions at nanoelectrodes, *Anal. Chem.* 78 (2006) 6526–6534. https://doi.org/10.1021/ac060924q.

[19] C.A. Nkuku, R.J. LeSuer, Electrochemistry in deep eutectic solvents, *J. Phys. Chem. B.* 111 (2007) 13271–13277. https://doi.org/10.1021/jp075794j.

[20] I.M. Kolthoff, A review of electrochemistry in non-aqueous solvents, *Pure Appl. Chem.* 25 (2009) 305–326. https://doi.org/10.1351/pac197125020305.

[21] A.M. Bond, M. Fleischmann, J. Robinson, Electrochemistry in organic solvents without supporting electrolyte using platinum microelectrodes, *J. Electroanal. Chem. Interfacial Electrochem.* 168 (1984) 299–312. https://doi.org/10.1016/0368-1874(84)87106-3.
[22] P.S. Toth, R.A.W. Dryfe, Novel organic solvents for electrochemistry at the liquid/liquid interface, *Analyst* 140 (2015) 1947–1954. https://doi.org/10.1039/c4an02250e.
[23] R.J. LeSuer, C. Buttolph, W.I. Geiger, Comparison of the conductivity properties of the tetrabutylammonium salt of tetrakis(pentafluorophenyl)borate anion with those of traditional supporting electrolyte anions in nonaqueous solvents, *Anal. Chem.* 76 (2004) 6395–6401. https://doi.org/10.1021/ac040087x.
[24] M. Ue, K. Ida, S. Mori, Electrochemical properties of organic liquid electrolytes based on quaternary onium salts for electrical double-layer capacitors, *J. Electrochem. Soc.* 141 (1994) 2989. https://doi.org/10.1149/1.2059270.
[25] W. Hyk, Z. Stojek, Generalized theory of steady-state voltammetry without a supporting electrolyte. Effect of product and substrate diffusion coefficient diversity, *Anal. Chem.* 74 (2002) 4805–4813. https://doi.org/10.1021/ac025765z.
[26] H.J. Walls, P.S. Fedkiw, T.A. Zawodzinski, S.A. Khan, Ion transport in silica nanocomposite electrolytes, *J. Electrochem. Soc.* 150 (2003) E165. https://doi.org/10.1149/1.1544635.
[27] F. Torres-Rojas, D. Muñoz, C. Pía Canales, S.A. Hevia, F. Leyton, N. Veloso, M. Isaacs, I.T. Vargas, Synergistic effect of electrotrophic perchlorate reducing microorganisms and chemically modified electrodes for enhancing bioelectrochemical perchlorate removal, *Environ. Res.* 233 (2023) 116442. https://doi.org/10.1016/j.envres.2023.116442.
[28] G.V. Lima, L. dos Santos Mello, E.R. Dockal, M. Firmino de Oliveira, Electrochemical determination of ethanol on carbon paste electrode chemically modified with [N,N′-cis-1,2-cyclohexylene bis (salicylideneaminate)] nickel(II) Schiff base complex, *Microchem. J.* 175 (2022) 107209. https://doi.org/10.1016/j.microc.2022.107209.
[29] A. Soares Castro, L. Silva de Azevedo, A. Thais Bruni, M. Manoel Teles de Menezes, E. Ralph Dockal, O. Vital de Oliveira, B. Ferreira, M. Firmino de Oliveira, Cocaine electrooxidation behavior, mechanism, and kinetics on a carbon paste electrode chemically modified with a cobalt or copper Schiff base complex, *Forensic Chem.* 26 (2021) 100363. https://doi.org/10.1016/j.forc.2021.100363.
[30] A. Wong, A.M. Santos, A.A. Cárdenas-Riojas, S.L. Calderon-Zavaleta, M.H.A. Feitosa, F.C. Moraes, A.M. Baena-Moncada, M.D.P.T. Sotomayor, Detection of 3,4-dichloroaniline in different matrices using chemically modified glassy carbon electrode, *Microchem. J.* 197 (2024) 109828. https://doi.org/10.1016/j.microc.2023.109828.
[31] A. Fujishima, K. Honda, Electrochemical photolysis of water at a semiconductor electrode, *Nature* 238 (1972) 37–38. https://doi.org/10.1038/238037a0.
[32] Y. Takatsuji, R. Wakabayashi, T. Sakakura, T. Haruyama, A “swingable” straight-chain affinity molecule immobilized on a semi-conductor electrode for photo-excited current-based molecular sensing, *Electrochim. Acta* 180 (2015) 202–207. https://doi.org/10.1016/j.electacta.2015.08.111.
[33] W. Chen, T. Jiao, Z. Diao, Z. Li, P. Chen, X. Dang, X. Dong, Y. Zhang, B. Zhang, Selective-area growth of β-Ga2O3 nanowire films on nano-patterned Si(111) substrate by metal-organic chemical vapor deposition, *Ceram. Int.* 49 (2023) 22170–22176. https://doi.org/10.1016/j.ceramint.2023.04.045.
[34] X. Yu, H. Li, W. Tian, Y. Ge, T. Wang, Z. Qi, J. Liu, Single-layer semiconductor-decorated flexible 2D protein nanosheets by engineered anchoring for efficient photocatalytic hydrogen production, *Int. J. Biol. Macromol.* 261 (2024) 129819. https://doi.org/10.1016/j.ijbiomac.2024.129819.
[35] J. Heinze, Ultramicroelectrodes in electrochemistry, *Angew. Chemie Int. Ed. English* 32 (1993) 1268–1288. https://doi.org/10.1002/anie.199312681.
[36] K. Aoki, Theory of ultramicroelectrodes, *Electroanalysis* 5 (1993) 627–639. https://doi.org/10.1002/elan.1140050802.

[37] J.C. Eklund, A.M. Bond, J.A. Alden, R.G. Compton, Perspectives in modern voltammetry: Basic concepts and mechanistic analysis, in: D. Bethell (Ed.), *Advances in Physical Organic Chemistry*, Academic Press, 1999: pp. 1–120. https://doi.org/10.1016/S0065-3160(08)60006-4.

[38] F.-R.F. Fan, A.J. Bard, Electrochemical detection of single molecules, *Science* 267 (1995) 871–874. https://doi.org/10.1126/science.267.5199.871.

[39] R.J. Forster, T.E. Keyes, Microelectrode applications, in: C.G. Zoski (Ed.), *Handbook of Electrochemistry*, Elsevier, Amsterdam, 2007: pp. 171–188. https://doi.org/10.1016/B978-044451958-0.50008-2.

[40] M.-L. Tercier, J. Buffle, In situ voltammetric measurements in natural waters: Future prospects and challenges, *Electroanalysis* 5 (1993) 187–200. https://doi.org/10.1002/elan.1140050303.

[41] A. Devadoss, J.D. Burgess, Steady-state detection of cholesterol contained in the plasma membrane of a single cell using lipid bilayer-modified microelectrodes incorporating cholesterol oxidase, *J. Am. Chem. Soc.* 126 (2004) 10214–10215. https://doi.org/10.1021/ja047856e.

[42] R.M. Wightman, P. Runnels, K. Troyer, Analysis of chemical dynamics in microenvironments, *Anal. Chim. Acta* 400 (1999) 5–12. https://doi.org/10.1016/S0003-2670(99)00605-4.

3 Electrocoagulation
Fundamentals and Mechanism

3.1 INTRODUCTION

A growing worldwide concern is the lack of fresh water. The World Economic Forum ranks the global water problem as the fourth most significant risk in terms of potential long-term effects on society. There is growing concern about the potential effects on human health, the environment, and the global economy as a result of a notable decline in the quality and quantity of fresh water accessible [1]. In addition, the United Nations (UN) ranked "Clean Water and Sanitation" as the sixth Sustainable Development Goal (SDG) to be accomplished by 2030 after acknowledging the significance and urgency of mitigating the freshwater problem. Thus, improving water quality, wastewater treatment, and safe reuse is Target 6.3 under Goal 6. Target 6.3, one of the eight targets to reach SDG goal 6, can be represented visually by dividing it down into the following four measurements according to UN definition [2]: Reducing pollution, eliminating, disposing of, and minimizing the release of chemicals and hazardous materials, halving the amount of untreated wastewater, and significantly expanding safe reuse and recycling worldwide are the goals. It is clear that urban societies and growing sectors are key factors boosting the global economy [3–9]. Disregarding the environmental effects of proper waste treatment and disposal, however, would eventually strain the global economy and contribute to the collapse of ecosystem sustainability, which may eventually have an impact on climate change. For example, Malaysia, one of the world's leading producers of palm oil, produced 19.86 million tons of crude palm oil (CPO) in 2019 alone—a 1.8% rise. A significant quantity of water is used for fresh fruit bunches (FFB) extraction procedures for each ton of CPO produced, with over 50% of the water being disposed of as effluents [10]. Furthermore, the effluents contain a high concentration of nontoxic organic matter and nutrients that have the potential to cause eutrophication and algal development over time, and conventional treatment techniques do not exceed Malaysia's Department of Environment's (DOE) environmental discharge criteria. Before being released into rivers, materials are often treated using chemical, biological, or physical techniques [2].

The various contaminants and wastewater kinds determine the sort of treatment that is needed. Advanced oxidation techniques [11], biological processes [12], physicochemical processes [13], and developing technologies like membrane filtration [14] and adsorption [15] are among the typical processes used in wastewater treatment. Strong oxidants are needed for advanced oxidation processes, which makes wastewater treatment expensive and dangerous. Conversely, biological processes necessitate tightly regulated environments with extended retention periods, more surface area,

DOI: 10.1201/9781003499671-3

and the production of undesired by-products. Chemical processes require a lot of chemical addition, which perplexes downstream operation by raising the possibility of secondary contamination. Wastewater cannot be effectively treated by membrane filtration and adsorption on their own; comprehensive pretreatment procedures must be included [2]. Alternatively, as a result of contaminants obstructing the pores and flux loss, the procedures eventually become unproductive. Therefore, electrochemical techniques have tremendous attraction in wastewater treatment research. Water treatment using electricity was initially proposed in the United Kingdom in 1889. Despite the fact that EC was successfully used in the United States in 1946 to treat significant amounts of drinking water, its broader applications were never widely adopted due to high capital costs and power supply constraints [16]. Nonetheless, ongoing advancements in EC research have amazingly maintained the importance of EC and its potentially beneficial effects in wastewater treatment. Due to its easy setup, compact size, and capacity to treat huge volumes of water without requiring substantial chemical treatment, electrochemical (EC) distillation stands out as the most environmentally friendly solution among all electrochemical methods used to treat wastewater [17]. Furthermore, because of its adaptability and configuration, the EC process is able to treat several types of wastewater from home and commercial sources that contain various kinds of pollutants. Until now, a large number of researchers from around the world have treated different kinds of wastewater using EC processes and have obtained encouraging results. To create cleaner effluents, several EC studies were advanced by merging with other processes and developing sophisticated wastewater systems [18].

This chapter examines the most recent developments on the evolution of EC, its reaction mechanism, and its characteristics. It also provides a summary of the essential principles of coagulation, which are required to comprehend the operation of EC and how its operational variables are affected.

3.2 COAGULATION PROCESSES

Since most wastewater contaminants are colloids, it is essential to have a basic understanding of colloidal chemistry in order to understand the idea of coagulation. Colloids dispersed in a solution, typically with sizes between 1 nm and 2 μm, are sufficiently durable to elude physical separation techniques including conventional filtration, flotation, and sedimentation. The balance of two important forces—the Van der Waal's attraction and the repulsive forces at the colloidal surface—determines the stability of the colloids. The colloids are particularly stable because of the strong repulsive force that is created by these charged layers working together [19–21]. Particle resistance to destabilization is one way to characterize the repulsive force, which is measured by comparing its potential differences. High colloidal stability inhibits the formation of flocs that sedimentation or precipitation can separate. The main goal of coagulation is to destabilize the colloidal surface charge so that the particles can group together to produce flocs that are simple to separate from water or wastewater. In summary, the primary goal of coagulation is to compress the pollutant's EDL, since the thickness of this layer indicates how strong the repulsion of the pollutant is [11, 22]. At optimal destabilization, the zeta potential approaches 0 mV and decreases as colloid destabilization increases. When counter ions outweigh the repulsive forces on the colloidal surface and adsorb on the colloidal surface, charge neutralization is also accomplished. As a result, the neutralized colloids congregate and group together in order to separate. Furthermore, metal coagulants can create dipole–dipole and hydrogen bonding interactions to create interparticle connections between colloids when they polymerize with the reactive groups. Colloidal instability results from the colloids' interparticle bridging growing in size. It should be remembered, too, that excessive polymer addition and quick mixing might lead to the dissociation of interparticle bridges and return colloid stability [23]. Conversely, when high concentrations of metal salts react with water, an insoluble sludge precipitate forms, trapping the contaminants. Therefore, it is possible to effectively study, control, and enhance the role of coagulation by having a solid grasp of the mechanisms that diminish colloidal stability.

3.2.1 Chemical Coagulation

Chemical coagulation (CC) is the process of adding coagulants to wastewater contaminants to destabilize their surface charges. Due to their availability, affordability, and high electrode dissolution rates, coagulants that have been extensively employed for effective pollutant removal are primarily the sulfates and chlorides of aluminum and iron [24]. Also, Al and Fe are often used for coagulation due to their enhanced capacity to produce polyvalent ions with strong coagulating properties. Metal hydroxides are formed after the coagulant dissociates in the solution, producing the polyvalent ions Al^{3+}, Fe^{2+}, and Fe^{3+} [25]. The charged hydroxyl ions combine with the contaminants in the solution to create complexes as the metal hydroxides form. As a result, the EDL compresses and collapses, allowing the pollutants to agglomerate and form flocs. Depending on their size and nature, these flocs either rise to the surface of the solution by flotation or sediment due to the effect of gravity to separate from the solution. It is feasible to treat wastewater with noticeable results using flocculation and CC. With the CC-flocculation technique, Sher et al. [19] obtained a remarkable 98% elimination of COD and 99% of turbidity. The main drawback of CC is chemical addition-induced secondary contamination. The removal of the harsh chemicals from the final treated water in CC requires additional separation procedures, which increases the complexity of the downstream operations. Consequently, the increasingly attractive method of electrochemical coagulation, which requires very little chemical input, comes into focus [26, 27].

3.2.2 Electrocoagulation

By applying electric current to the pollutants, EC is an electrochemical technique that destabilizes their charges, causing electrode disintegration and trapping the pollutants in flocs that can be removed from the electrolytic mixture. The dissolved anode and subsequent hydrolysis produce metal ions that function as coagulants, weakening charged contaminants and causing them to clump together. Because of their polar moment and capacity to oxidize, the metal hydroxyl ions combine with the contaminants to create complexes that cause floc formation. For the purpose of treating wastewater, an insulated container holds electrodes and a power source [28, 29]. Figure 3.1 depicts a schematic diagram of a basic EC reactor connection. The electrode type, electrode spacing, applied current density, starting pH, electrolyte conductivity, and treatment duration are the main variables that impact the electrochemical process. For this reason, every EC study needs to optimize the operational variables.

When current is applied, these sacrificial metal electrodes release metal hydroxides into the electrolyte, which have a strong propensity to interact with the contaminants in the electrolytic cell [30]. Faraday's law is used to estimate the metal ions that are liberated during electrode dissociation as EC reactions progress. It is critical to emphasize the aspects of EC that noticeably surpass CC. Because EC operates without the need for moving parts, it is simpler to use and requires less maintenance. Aswathy et al. (2016) demonstrated a markedly different COD elimination percentage when comparing the efficacy of EC and CC in the treatment of bilge water, which is the oily wastewater generated by oceangoing vessels [31]. 78% of COD was removed by EC as opposed to 59% by CC. EC had a bigger floc size and less bound water, making it possible to separate it fast through filtration. Furthermore, as

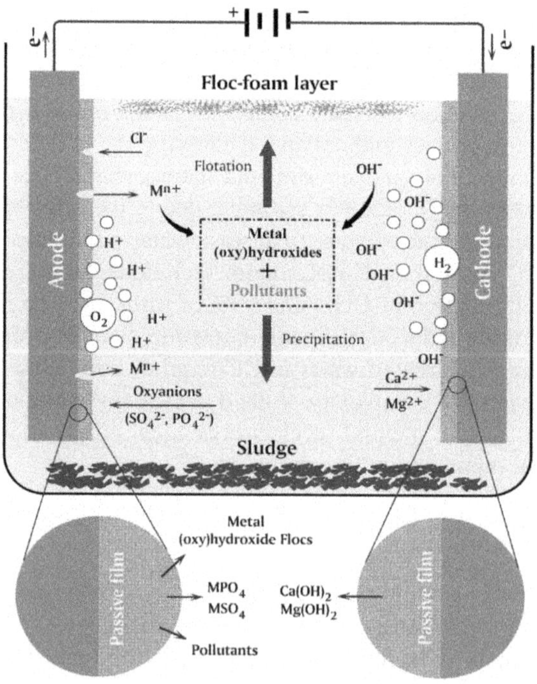

FIGURE 3.1 Schematic representation of an EC reactor (Reproduced with permission from Abdollahi et al. (2023) [34]).

the production of metal coagulants is powered by electricity, less chemical addition is required in EC. Thus, using solar power, some researchers examined EC-based wastewater treatment and obtained extremely encouraging findings. Therefore, EC can further support wastewater treatment's environmental and economic sustainability [32, 33].

3.3 REACTION MECHANISM AND DEGRADATION KINETICS OF THE ELECTROCOAGULATION PROCESS

The EC mechanism consists of seven key steps: (i) production of a metal cation by passing current through anodes; (ii) generation of a hydroxyl ion as a result of hydrolysis of the cathode; (iii) the interaction between the metal cation and the hydroxyl ion forming metal hydroxide; (iv) conversion of hazardous pollutants into safe intermediary substances; (v) contaminant's charge neutralization due to its interaction with the metal hydroxide; (vi) the metal hydroxide subsequently adsorbs the charge-neutralized pollutants, which are then removed by sweep coagulation; and (vii) production of H_2 gas (cathode) carries the flocs upward by sweep flotation [5, 28].

Figure 3.2 illustrates the procedure for removing pollutants by electrocoagulation. Anodic dissolution (Eq. (3.1)), H_2 gas formation (cathode) (Eq. (3.2)), and hydroxyl ion formation (anode) (Eq. (3.3)), all result in the generation of *in situ* flocs in the EC cell. The floc that is surrounded by metal hydroxides and acts as an effective adsorbent is the result of these electrogenerated coagulants taking place [17].

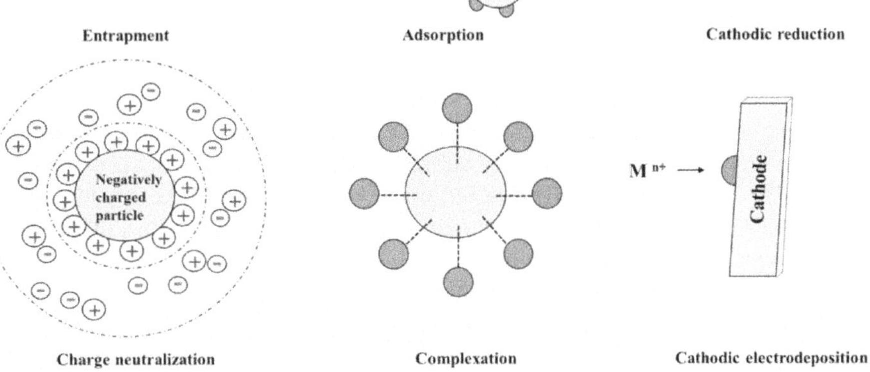

FIGURE 3.2 Contaminant degradation mechanism through the electrocoagulation method (Reproduced with permission from Das et al. (2022) [5]).

$$\text{At anode: M(s)} \rightarrow \text{M}^{3+}_{\text{aq}} + \text{ne}^- \quad (3.1)$$

$$\text{H}_2\text{O} \rightarrow 4\text{H}^+_{\text{aq}} + \text{O}_2\text{(g)+4e}^- \quad (3.2)$$

$$\text{At cathode: nH}_2\text{O + ne}^- \rightarrow \frac{n}{2}\text{H2(g)} + \text{nOH}^-_{\text{aq}} \quad (3.3)$$

The pollutant removal kinetics of the electrocoagulation process have been extensively studied during the past 20 years. The kinetic characteristics of the batch electrocoagulation method for the elimination of suspended particles and heavy metals from synthetic wastewater were ascertained by Singh and Mishra (2017) [35]. After a treatment duration of 30 minutes with an applied current density of 5.3 mA/cm^2, the study showed an elimination rate of 78.8% for total suspended solids (TSS) and the removal efficiencies of 61%, 83%, and 98% for the heavy metals chromium, zinc, and copper, respectively. El-Shazly and Daous (2013) assessed the phosphate removal kinetics from industrial wastewater using the continuous EC procedure [36]. The impact of various operational parameters on the rates of phosphate removal was thoroughly investigated in this study. The rate of phosphate depletion decreases as the concentration of phosphate in the solution increases, while the rate of phosphate elimination increases as the temperature, flow rate, and current density increase. At greater starting phosphate concentrations, a decrease in process efficiency was noted. According to the study, diffusion served as the primary reaction driving force, and the first-order reaction kinetics best explained the phosphate removal kinetic behavior. The kinetic analysis successfully reveals the working of chemical reactions, establishes time-dependent predictions about the make-up of reaction mixtures, and pinpoints the phenomenon of rate control.

3.4 FACTORS INFLUENCING THE ELECTROCOAGULATION PROCESS

3.4.1 Current Density

The kind of pollutants in the effluent are mostly responsible for the variations in ionic interaction that generate this fluctuation in current density. The range of applied current density in the EC process is often 0.01 to 880 A/m^2. Numerous studies have investigated how the applied current density affects various wastewaters that contain pollutants [6, 37]. Eryuruk et al. (2018) treated poultry slaughterhouse wastewater polluted with high COD concentrations (9000 mg/L) using electrocoagulation [38]. The process performance was assessed in relation to variations in current density (30–50 mA/cm^2). Within 40 minutes of treatment, the maximum COD elimination effectiveness of 96% was achieved at an ideal current density of 50 mA/cm^2. It was determined that by optimizing the current density, the EC process could successfully treat wastewaters with high levels of COD contamination. It is generally known that increased pollutant removal efficiency is a function of increasing current density. However, excess current density may lead to the production of *in situ* secondary reactions and colloidal charge reversal. This could lead to the formation of a thin-film layer on the electrode surface and quicker electrode depletion, which would significantly reduce process efficiency. Because of this, the targeted effluents should determine how best to optimize the current density. The aforementioned difficulties can also be addressed using a larger electrode surface area at a lower or optimum current density [39].

3.4.2 Inter-electrode Distance

The inter-electrode spacing, which controls the electrostatic field between the cathodes and anodes, is regarded as a highly important parameter in the EC process. The electrostatic field gradually rises as the inter-electrode distance decreases. Because of the strong collisions brought on by the strong electrostatic attraction, the metal hydroxides that help form flocs to encourage coagulation degrade. As a result, the EC process becomes much more efficient at shorter inter-electrode distances [5, 40]. Furthermore, because of the low electrostatic forces, a wider inter-electrode distance postpones the following formation of flocs based on metal hydroxides. In addition to decreasing process efficiency, an excessive electrode distance leads to high power consumption as a means of making up for the delayed release of metal ions between the cathodes and anodes. Operating the EC process at the ideal inter-electrode spacing is therefore crucial. According to several studies, the ideal inter-electrode distance for various wastewaters is between 0.5 and 1 cm [30]. A smaller electrode gap of 0.5 cm has been shown by Rahman et al. (2020) to greatly improve the efficiency of the electrochemical process. Using galvanized Fe electrodes, the study demonstrated an efficient removal of 94.5% turbidity, 86.2% color, and 25% TOC from peat water [41]. Inter-electrode distance was examined by Akarsu et al. (2016) [42] and Aswathy et al. (2016) [31] in relation to the treatment of bilge wastewater. In contrast to Aswathy et al. (2016), who reported an 89.9% removal efficiency in 120 minutes, Akarsu et al. (2016) reported a COD removal effectiveness of 52% within 90 minutes of treatment time. For both trials, the main deciding factors were revealed to be the inter-electrode distance and treatment time. The primary cause of the significant variance in removal effectiveness is the inter-electrode distance. Akarsu et al. (2016) used an electrode spacing of 8 cm, while Aswathy et al. (2016) used an inter-electrode spacing of 1 cm. Mousazadeh et al. (2021) examined the utilization of the EC procedure to remediate wastewater with elevated COD content [43]. It has been demonstrated that a reduction in the distance between electrodes greatly enhances process efficiency. Nevertheless, at larger inter-electrode separation, ohmic loss relative to the cathode and the anode over voltage increases, slowing the synthesis of coagulants.

3.4.3 Electrode Configuration

Bipolar electrodes in serial connection (BP-S) and monopolar electrodes in serial (MP-S) and parallel connection (MP-P) are the most widely utilized electrode configurations. Figure 3.3 illustrates how different electrode connection options are arranged in an EC reactor. Each electrode in a monopolar configuration mode serves as either the cathode or the anode in the electrochemical cell depending on the electrical polarity. When

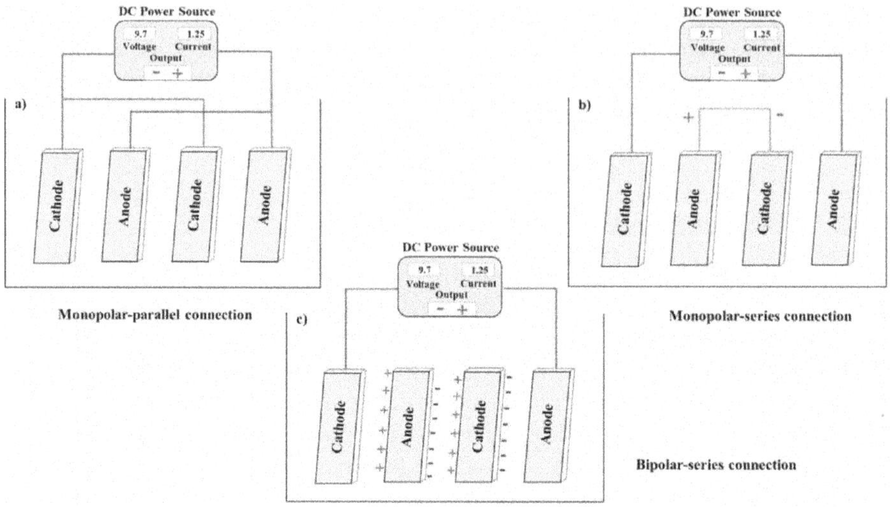

FIGURE 3.3 Various electrode connections in an EC reactor: (a) monopolar-parallel, (b) monopolar-series, and (c) bipolar series (Reproduced with permission from Das et al. (2022) [5]).

there is an MP-P connection, the EC cell's sacrificial anodes are directly connected to one another and the cathodes are likewise connected in a comparable fashion [17, 44]. Each anode–cathode pair in an MP-S connection is internally connected; the outer electrodes are not connected to them. It is important to remember that a series connection requires a larger potential difference even though the current flows through each electrode at the same rate. The parallel connection mode offers substantial energy-saving advantages. Additionally, the MP-P connection offers minimal operating costs, and the BP-S connection mode assists in high pollution removal [45]. The efficacy of bipolar and monopolar electrode configurations for the removal of fluoride from drinking water was examined in a study by Ghosh et al. (2008) [46]. Bipolar connection mode was found to have a higher fluoride removal effectiveness (85%) and require shorter treatment time (45 minutes) than monopolar connection mode. It was concluded that although bipolar connections offer higher pollution removal effectiveness, monopolar connections are more economical. A bipolar electrode configuration, which involves two electrochemical cells working together, has a larger surface area than a monopolar configuration among the electrode connection options. This sufficiently supports the bipolar mode anodic oxidation, which improves the EC process's efficiency. However, because there is less surface area available in the monopolar electrode layout, the efficiency is significantly lower even with the same applied current density for both configurations. On the other hand, minimal running costs are seen when connecting the monopolar electrodes [5].

3.4.4 Initial pH

The effectiveness of the EC process is significantly influenced by pH. The diversity and abundance of these complexes are crucial because they all show unique interactions with pollutants in wastewater, leading to varying coagulant capabilities. $FE(OH)_3^-$ and $AL(OH)_3^-$ that develop in neutral or slightly alkaline/acidic conditions greatly increase the coagulants' ability to adsorb. Hence, the double layer is changed, which has an impact on the development of the coagulant. The pH of the solution must therefore be adjusted in accordance with the various physicochemical features of the targeted contaminants and the effluent properties [47, 48]. Bener et al. (2019) examined the impact of three distinct pH values—5, 8, and 10—on the effectiveness of TOC removal from actual textile effluent using Al electrodes [49]. At the pH of 5, the highest removal of TOC (65%) was achieved in the solution. Additionally, it was found that up to a pH of 10, increasing the solution's pH causes the removal efficiency to decline. Two variables are responsible for the decline in EC removal effectiveness at pH 10: (a) the species $AL(OH)_4^-$, which occurs under extremely alkaline conditions for Al electrodes, has a low adsorption efficiency; and (b) the production of H_2 gas near the cathode. Irdemez et al. (2006) looked at the

TABLE 3.1
Performance efficiency of the EC method in different types of effluents (Reproduced with permission from Tahreen et al. (2020) [2]).

Effluent type	Electrode combination	Electrode distance	Current density	Initial pH	EC time	% Pollutant removal
Textile effluent	Al-Al	20mm	25 mA/cm^2	5	120 min	18.6% COD and 83.5% turbidity
Laundry effluent	Graphite-Steel	10mm	5.26 mA/cm^2	5.5	5 min	90% surfactant
Oily effluent	Al-Al	5mm	8 mA/cm^2	3.6	20 min	71% oil & grease and 14% TDS
Municipal effluent	Al-Al	10mm	4 mA/cm^2	7.4–8.5	20 min	92% COD, and 93.9% turbidity
Palm oil mill effluent	Al-Al	30mm	40.21 mA/cm^2	4.4	45.67 min	71.3% COD and 96.8% color
Textile effluent	Al-Al	10mm	40 mA/cm^2	7	30min	95% color and 99% turbidity
Urban effluent	Al-Al	30mm	20 mA/cm^2	7.4	30 min	85% COD 94% TSS
Textile effluent	Fe-Al	30mm	2 mA/cm^2	8	80 min	90% COD 90% color
Bio-treated municipal effluent	Al-Al	5mm	0.265	7.5	40 min	63% NO_3^- and 79.9% SO_4^{2-}
Bilge effluent	Al-Al	80mm	10 mA/cm^2	8	90 min	52% COD

impact of different pH levels on the removal of phosphate from synthetic wastewater using Fe electrodes in a different investigation [50]. It was noted that at a lower pH of 3.4, a significant decrease in the energy usage of the EC method was seen. Additionally, it was shown that the pH range of 6–7 was ideal for the best phosphate removal efficiency (93%) due to the lowest solubility of Fe $(OH)_2$. During the EC process, the release and production of coagulants are significantly regulated by the pH of the solution. The pH of the solution is significantly raised by the development of H_2 near the cathode and the creation of OH^- ions. Inappropriate pH levels, on the other hand, could change the double layer and lessen the generation of coagulants. Moreover, the adsorption capacity of the monomeric and polymeric species generated under extremely alkaline conditions is extremely poor. Therefore, it is important to efficiently maintain the pH variation in order to enhance the coagulant species' ability to adsorb pollutants during the EC process. Table 3.1 illustrates how various operating parameters affect the EC process's removal efficiency.

3.5 ELECTRODE PASSIVATION

Reliable EC performance requires maintaining a high and constant faradaic efficiency over an extended period of operation. Electrode passivation can be influenced by a number of variables, including pH, composition, current density, waveform, reactor design, dissolved oxygen content, pollutant concentration, and electrode material. The composition of the solution is one of the most crucial of these variables. Due to the wide range of pollutant species present in different water and wastewater matrices, the passivation rate might fluctuate significantly. Furthermore, passivation has been observed to be exacerbated by natural contaminants found in water, such as natural organic matter (NOM) [51].

3.5.1 Passivation Mechanism

Electrochemical and physicochemical processes involving coagulants and aqueous-phase species interacting and precipitating on the electrode surface give rise to electrode passivation. Around the anode and the cathode, respectively, water electrolysis creates pH conditions that are acidic and alkaline. The primary processes and reactions that lead to the development of the surface layer and improved passivation are as follows: (1) A portion of metal (oxy)hydroxide flocs precipitate on the electrode surface; (2) oxyanions bond to metal ions to form the oxyanion-rich passivating film on the anode surface; (3) minerals like calcium and magnesium precipitate on the cathode surface; (4) contaminant residues accumulate on the electrode surface; and (5) other particles in solution are adsorbed to the anode or cathode surface [34, 52].

3.5.2 Adverse Effects of Passivation

It is anticipated that the surface layer on the electrode surface will thicken with time because there is a lower chance of metal ions migrating through the layer and into the bulk solution. Solid accumulation can cause a number of operational issues if the surface layer is made of materials that conduct poorly. Pitting corrosion of the passivated electrode can lead to structural failure and shorter periods between periodic replacements. Additionally, electrode passivation obstructs normal electron flow and renders the electrode surface quasi-inert [53]. Ohmic resistance and over potential are consequently increased by this process, leading to excessive energy consumption and increased operational expenses. Furthermore, passive film production at high enough anodic over potentials promotes water oxidation over electrode oxidation, which lowers the quantity of produced flocs and significantly lowers pollutant removal effectiveness. Furthermore, the system may shut down as a result of extreme passivation [54].

3.5.3 Mitigation Approaches

Electrode passivation is required to sustain the electrical conductivity (EC) process for prolonged periods of time, as the development and expansion of the passivating layer on the electrode surface may have adverse effects. The following lists the principal techniques used in EC to de-passivate electrodes, along with some advantages and disadvantages of each [55]:

(1) Cleaning electrodes by mechanical or chemical means. Benefits: efficient de-passivation and no need for additional tools. Drawbacks: high personnel costs; system shutdown is necessary (which makes automating EC difficult); and acid waste management is necessary [56].
(2) By altering the electrode shape, reactor geometry, or flow velocity, the turbulent state can be increased. Benefits: no chemicals are required and *in situ* de-passivation is successful. Drawbacks: needs a continuous pump and not suitable for batch systems [57].
(3) Putting in a strong field of gravity. Benefits: no chemicals are required and *in situ* de-passivation is successful. Drawbacks: complicated setup and additional infrastructure needed [58].
(4) EC and ultrasonic processes combined (Sono-EC). Benefits: no chemicals are required and *in situ* de-passivation is successful. Drawbacks: costly, needs specialized infrastructure, and could lead to breaking down of flocs [59].
(5) Using waveforms for current other than DC. Benefits: complete automatability, efficient *in situ* de-passivation, no need for chemicals, no complicated design modifications, and no additional treatment stages. Drawbacks: need for additional electrical equipment [60].

While strategy 1 through 4 may be useful in mitigating passivation, they necessitate the addition of chemicals, the termination of systems, or the installation of specialized infrastructure, all of which significantly increase the process's cost and complexity. Nonetheless, due to the numerous benefits mentioned above, EC operation with current waveforms other than DC (strategy 5) has grown in popularity in recent years. This technique is also more preferable because of its cost-effectiveness, lesser sludge output, homogeneous electrode corrosion, and decreased energy and electrode usage [34, 61]. Consequently, the present review centers on this approach and assesses its efficacy in electrode de-passivation and EC performance augmentation.

3.6 CONCLUSION

Many published works have amply established the validity of the EC process for the removal of wastewater contaminants at both lab and pilot scale operations. Although the physicochemical properties of wastewater play a significant role in its removal mechanism, the EC process is nevertheless regarded as extremely efficient when compared to other conventional procedures. Higher pollutant removal efficiency can be achieved using Al or Fe electrodes; however, total removal is not always achieved. Numerous operating factors such as current density, electrode type, pH, time, electrolyte, and electrode gap mostly determine the efficiency of the EC process. The development of new electrode materials, various electrode connection methods, ideal reactor designs, simulation models, and other factors may all significantly contribute to improving the EC process's treatment effectiveness. Furthermore, the pollutant removal efficacy of the EC process is increased when combined with other water treatment methods. Additionally, it was noted that the hybrid EC procedure significantly lowers the treatment's overall energy consumption and operating costs. Because of its minimal environmental impact, simple operation handling, and ease of use, the hybrid technology that combines membrane filtration with EC has been shown to be one of the most promising methods for treating wastewater and water. It was concluded that using EC as a pretreatment procedure has several benefits, including reduction of the total treatment cost of the hybrid process, improvement of permeate water quality, and reduction of membrane fouling.

REFERENCES

[1] M. Sharma, P.P. Das, T. Sood, A. Chakraborty, M.K. Purkait, Ameliorated polyvinylidene fluoride based proton exchange membrane impregnated with graphene oxide, and cellulose acetate obtained from sugarcane bagasse for application in microbial fuel cell, *J. Environ. Chem. Eng.* 9 (2021) 106681. https://doi.org/10.1016/j.jece.2021.106681.

[2] A. Tahreen, M.S. Jami, F. Ali, Role of electrocoagulation in wastewater treatment: A developmental review, *J. Water Process Eng.* 37 (2020) 101440. https://doi.org/10.1016/j.jwpe.2020.101440.

[3] P.P. Das, P. Mondal, Anweshan, A. Sinha, P. Biswas, S. Sarkar, M.K. Purkait, Treatment of steel plant generated biological oxidation treated (BOT) wastewater by hybrid process, *Sep. Purif. Technol.* 258 (2021). https://doi.org/10.1016/j.seppur.2020.118013.

[4] P.P. Das, A. Anweshan, M.K. Purkait, Treatment of cold rolling mill (CRM) effluent of steel industry, *Sep. Purif. Technol.* 274 (2021) 119083. https://doi.org/10.1016/j.seppur.2021.119083.

[5] P.P. Das, M. Sharma, M.K. Purkait, Recent progress on electrocoagulation process for wastewater treatment: A review, *Sep. Purif. Technol.* 292 (2022) 121058. https://doi.org/10.1016/j.seppur.2022.121058.

[6] P.P. Das, A. Anweshan, P. Mondal, A. Sinha, P. Biswas, S. Sarkar, M.K. Purkait, Integrated ozonation assisted electrocoagulation process for the removal of cyanide from steel industry wastewater, *Chemosphere* 263 (2021) 128370. https://doi.org/10.1016/j.chemosphere.2020.128370.

[7] N.S. Samanta, P.P. Das, P. Mondal, M. Changmai, M.K. Purkait, Critical review on the synthesis and advancement of industrial and biomass waste-based zeolites and their applications in gas adsorption and biomedical studies, *J. Indian Chem. Soc.* 99 (2022) 100761. https://doi.org/10.1016/j.jics.2022.100761.

[8] M. Sharma, P.P. Das, A. Chakraborty, M.K. Purkait, Clean energy from salinity gradients using pressure retarded osmosis and reverse electrodialysis: A review, *Sustain. Energy Technol. Assessments* 49 (2022) 101687. https://doi.org/10.1016/j.seta.2021.101687.
[9] A.D. Sontakke, P.P. Das, P. Mondal, M.K. Purkait, Thin-film composite nanofiltration hollow fiber membranes toward textile industry effluent treatment and environmental remediation applications: Review, *Emergent Mater.* (2021). https://doi.org/10.1007/s42247-021-00261-y.
[10] A.L. Ahmad, S. Ismail, S. Bhatia, Water recycling from palm oil mill effluent (POME) using membrane technology, *Desalination* 157 (2003) 87–95. https://doi.org/10.1016/S0011-9164(03)00387-4.
[11] P.P. Das, S. Dhara, M.K. Purkait, Ozone-based oxidation processes for the removal of pharmaceutical products from wastewater, in: M.P. Shah, P. Ghosh (Eds.), *Development in Wastewater Treatment Research and Processes*, Elsevier, 2024: pp. 287–308. https://doi.org/10.1016/B978-0-443-19207-4.00003-3.
[12] M. Sharma, P.P. Das, T. Sood, A. Chakraborty, M.K. Purkait, Reduced graphene oxide incorporated polyvinylidene fluoride/cellulose acetate proton exchange membrane for energy extraction using microbial fuel cells, *J. Electroanal. Chem.* 907 (2022) 115890. https://doi.org/10.1016/j.jelechem.2021.115890.
[13] N.S. Samanta, P.P. Das, P. Mondal, U. Bora, M.K. Purkait, Physico-chemical and adsorption study of hydrothermally treated zeolite A and FAU-type zeolite X prepared from LD (Linz—Donawitz) slag of the steel industry, *Int. J. Environ. Anal. Chem.* (n.d.) 1–23. https://doi.org/10.1080/03067319.2022.2079082.
[14] M. Changmai, P.P. Das, P. Mondal, M. Pasawan, A. Sinha, P. Biswas, S. Sarkar, M.K. Purkait, Hybrid electrocoagulation—microfiltration technique for treatment of nanofiltration rejected steel industry effluent, *Int. J. Environ. Anal. Chem.* 102 (2022) 62–83. https://doi.org/10.1080/03067319.2020.1715381.
[15] N. Shekhar Samanta, P.P. Das, S. Dhara, M.K. Purkait, An overview of precious metal recovery from steel industry slag: Recovery strategy and utilization, *Ind. Eng. Chem. Res.* 62 (2023) 9006–9031. https://doi.org/10.1021/acs.iecr.3c00604.
[16] G. Chen, Electrochemical technologies in wastewater treatment, *Sep. Purif. Technol.* 38 (2004) 11–41. https://doi.org/10.1016/j.seppur.2003.10.006.
[17] P.P. Das, A.D. Sontakke, M.K. Purkait, Electrocoagulation process for wastewater treatment: Applications, challenges, and prospects, in: M.P. Shah (Ed.), *Development in Wastewater Treatment Research and Processes*, Elsevier, 2023: pp. 23–48. https://doi.org/10.1016/B978-0-323-95684-0.00015-4.
[18] P.P. Das, S. Dhara, M.K. Purkait, Hybrid electrocoagulation and ozonation techniques for industrial wastewater treatment, in: M.P. Shah (Ed.), *Sustainable Industrial Wastewater Treatment and Pollution Control*, Springer Nature Singapore, Singapore, 2023: pp. 107–128. https://doi.org/10.1007/978-981-99-2560-5_6.
[19] A.S. Naje, S. Chelliapan, Z. Zakaria, M.A. Ajeel, P.A. Alaba, A review of electrocoagulation technology for the treatment of textile wastewater, *Rev. Chem. Eng.* 33 (2017) 263–292. https://doi.org/10.1515/revce-2016-0019.
[20] M. Sharma, P.P. Das, M.K. Purkait, Energy storage properties of nanomaterials, in: A. Husen, K.S. Siddiqi (Eds.), *Advances in Smart Nanomaterials and their Applications*, Elsevier, 2023: pp. 337–350. https://doi.org/10.1016/B978-0-323-99546-7.00005-7.
[21] M. Sharma, P.P. Das, A. Chakraborty, M.K. Purkait, Extraction of clean energy from industrial wastewater using bioelectrochemical process, in: M. Sillanpää, A. Khadir, K. Gurung (Eds.), *Resource Recovery in Industrial Waste Waters*, Elsevier, 2023: pp. 601–620. https://doi.org/10.1016/B978-0-323-95327-6.00003-8.
[22] P.P. Das, Deepti, M.K. Purkait, Industrial wastewater to biohydrogen production via potential bio-refinery route, in: M.P. Shah (Ed.), *Biorefinery for Water and Wastewater Treatment*, Springer International Publishing, Cham, 2023: pp. 159–179. https://doi.org/10.1007/978-3-031-20822-5_8.

[23] B.A. Firth, R.J. Hunter, Flow properties of coagulated colloidal suspensions: I. Energy dissipation in the flow units, *J. Colloid Interface Sci.* 57 (1976) 248–256. https://doi.org/10.1016/0021-9797(76)90200-9.
[24] O.T. Can, E. Gengec, M. Kobya, TOC and COD removal from instant coffee and coffee products production wastewater by chemical coagulation assisted electrooxidation, *J. Water Process Eng.* 28 (2019) 28–35. https://doi.org/10.1016/j.jwpe.2019.01.002.
[25] A.K. Verma, R.R. Dash, P. Bhunia, A review on chemical coagulation/flocculation technologies for removal of colour from textile wastewaters, *J. Environ. Manage.* 93 (2012) 154–168. https://doi.org/10.1016/j.jenvman.2011.09.012.
[26] M. Sharma, P.P. Das, S. Kumar, M.K. Purkait, Polyurethane foams as packing and insulating materials, in: *Polyurethanes: Preparation, Properties, and Applications*, Vol. *Emerging Applications*, American Chemical Society, 2023: pp. 4–83. https://doi.org/10.1021/bk-2023-1454.ch004.
[27] P.P. Das, P. Mondal, Membrane-assisted potable water reuses applications: Benefits and drawbacks, in: M. Sillanpää, A. Khadir, K. Gurung (Eds.), *Resource Recovery in Drinking Water Treatment*, Elsevier, 2023: pp. 289–309. https://doi.org/10.1016/B978-0-323-99344-9.00014-1.
[28] M. Bharti, P.P. Das, M.K. Purkait, A review on the treatment of water and wastewater by electrocoagulation process: Advances and emerging applications, *J. Environ. Chem. Eng.* 11 (2023) 111558. https://doi.org/10.1016/j.jece.2023.111558.
[29] S. Dhara, N. Shekhar Samanta, P.P. Das, R.V.S. Uppaluri, M.K. Purkait, Ravenna grass-extracted alkaline lignin-based polysulfone mixed matrix membrane (MMM) for Aqueous Cr(VI) removal, *ACS Appl. Polym. Mater.* 5 (2023) 6399–6411. https://doi.org/10.1021/acsapm.3c00999.
[30] O. Sahu, B. Mazumdar, P.K. Chaudhari, Treatment of wastewater by electrocoagulation: A review, *Environ. Sci. Pollut. Res.* 21 (2014) 2397–2413. https://doi.org/10.1007/s11356-013-2208-6.
[31] P. Aswathy, R. Gandhimathi, S.T. Ramesh, P.V. Nidheesh, Removal of organics from bilge water by batch electrocoagulation process, *Sep. Purif. Technol.* 159 (2016) 108–115. https://doi.org/10.1016/j.seppur.2016.01.001.
[32] D. Ghosh, H. Solanki, M.K. Purkait, Removal of Fe(II) from tap water by electrocoagulation technique, *J. Hazard. Mater.* 155 (2008) 135–143. https://doi.org/10.1016/j.jhazmat.2007.11.042.
[33] P.P. Das, S. Dhara, M.K. Purkait, The anaerobic ammonium oxidation process: Inhibition, challenges and opportunities, in: M.P. Shah (Ed.), *Ammonia Oxidizing BacteriaApplications in Industrial Wastewater Treatment*, Royal Society of Chemistry, 2023: pp. 56–82. https://doi.org/10.1039/BK9781837671960-00056.
[34] J. Abdollahi, M.R. Alavi Moghaddam, S. Habibzadeh, The role of the current waveform in mitigating passivation and enhancing electrocoagulation performance: A critical review, *Chemosphere* 312 (2023) 137212. https://doi.org/10.1016/j.chemosphere.2022.137212.
[35] H. Singh, B.K. Mishra, Assessment of kinetics behavior of electrocoagulation process for the removal of suspended solids and metals from synthetic water, *Environ. Eng. Res.* 22 (2017) 141–148. https://doi.org/10.4491/eer.2016.029.
[36] A.H. El-Shazly, M.A. Daous, Kinetics and performance of phosphate removal from hot industrial effluents using a continuous flow electrocoagulation reactor, *Int. J. Electrochem. Sci.* 8 (2013) 184–194.
[37] P.P. Das, N.S. Samanta, S. Dhara, M.K. Purkait, Biofuel production from algal biomass, in: M.P. Shah (Ed.), *Green Approach to Alternative Fuel for a Sustainable Future*, Elsevier, 2023: pp. 167–179. https://doi.org/10.1016/B978-0-12-824318-3.00009-6.
[38] K. Eryuruk, U. Tezcan Un, U. Bakır Ogutveren, Electrochemical treatment of wastewaters from poultry slaughtering and processing by using iron electrodes, *J. Clean. Prod.* 172 (2018) 1089–1095. https://doi.org/10.1016/j.jclepro.2017.10.254.

[39] M. Bayramoglu, M. Kobya, M. Eyvaz, E. Senturk, Technical and economic analysis of electrocoagulation for the treatment of poultry slaughterhouse wastewater, *Sep. Purif. Technol.* 51 (2006) 404–408. https://doi.org/10.1016/j.seppur.2006.03.003.
[40] P.P. Das, A.D. Sontakke, M.K. Purkait, Rice straw for biofuel production, in: M.P. Shah (Ed.), *Green Approach to Alternative Fuel for a Sustainable Future*, Elsevier, 2023: pp. 153–166. https://doi.org/10.1016/B978-0-12-824318-3.00034-5.
[41] N.A. Rahman, N.A. Tomiran, A.H. Hashim, Batch electrocoagulation treatment of peat water in Sarawak with galvanized iron electrodes, *Mater. Sci. Forum* 997 MSF (2020) 127–138. https://doi.org/10.4028/www.scientific.net/MSF.997.127.
[42] C. Akarsu, Y. Ozay, N. Dizge, H. Elif Gulsen, H. Ates, B. Gozmen, M. Turabik, Electrocoagulation and nanofiltration integrated process application in purification of bilge water using response surface methodology, *Water Sci. Technol.* 74 (2016) 564–579. https://doi.org/10.2166/wst.2016.168.
[43] M. Mousazadeh, E.K. Niaragh, M. Usman, S.U. Khan, M.A. Sandoval, Z. Al-Qodah, Z. Bin Khalid, V. Gilhotra, M.M. Emamjomeh, A critical review of state-of-the-art electrocoagulation technique applied to COD-rich industrial wastewaters, *Environ. Sci. Pollut. Res.* 28 (2021) 43143–43172. https://doi.org/10.1007/s11356-021-14631-w.
[44] S. Chakraborty, P.P. Das, P. Mondal, Recent advances in membrane technology for the recovery and reuse of valuable resources, in: M. Sillanpää, A. Khadir, K. Gurung (Eds.), *Resource Recovery in Industrial Waste Waters*, Elsevier, 2023: pp. 695–719. https://doi.org/10.1016/B978-0-323-95327-6.00028-2.
[45] M. Kobya, M. Bayramoglu, M. Eyvaz, Techno-economical evaluation of electrocoagulation for the textile wastewater using different electrode connections, *J. Hazard. Mater.* 148 (2007) 311–318. https://doi.org/10.1016/j.jhazmat.2007.02.036.
[46] D. Ghosh, C.R. Medhi, M.K. Purkait, Treatment of fluoride containing drinking water by electrocoagulation using monopolar and bipolar electrode connections, *Chemosphere* 73 (2008) 1393–1400. https://doi.org/10.1016/j.chemosphere.2008.08.041.
[47] M. Kobya, E. Demirbas, M. Sözbir, Decolorisation of aqueous reactive dye Remazol Red 3B by electrocoagulation, *Color. Technol.* 126 (2010) 282–288. https://doi.org/10.1111/j.1478-4408.2010.00259.x.
[48] P.P. Das, P. Mondal, M.K. Purkait, Recent advances in synthesis of iron nanoparticles via green route and their application in biofuel production, in: M. Srivastava, M.A. Malik, P.K. Mishra (Eds.), *Green Nano Solution for Bioenergy Production Enhancement*, Springer Nature Singapore, Singapore, 2022: pp. 79–104. https://doi.org/10.1007/978-981-16-9356-4_4.
[49] S. Bener, Ö. Bulca, B. Palas, G. Tekin, S. Atalay, G. Ersöz, Electrocoagulation process for the treatment of real textile wastewater: Effect of operative conditions on the organic carbon removal and kinetic study, *Process Saf. Environ. Prot.* 129 (2019) 47–54. https://doi.org/10.1016/j.psep.2019.06.010.
[50] Ş. Irdemez, N. Demircioğlu, Y.Ş. Yildiz, The effects of pH on phosphate removal from wastewater by electrocoagulation with iron plate electrodes, *J. Hazard. Mater.* 137 (2006) 1231–1235. https://doi.org/10.1016/j.jhazmat.2006.04.019.
[51] E. Mohora, S. Rončević, J. Agbaba, A. Tubić, M. Mitić, M. Klašnja, B. Dalmacija, Removal of arsenic from groundwater rich in natural organic matter (NOM) by continuous electrocoagulation/flocculation (ECF), *Sep. Purif. Technol.* 136 (2014) 150–156. https://doi.org/10.1016/j.seppur.2014.09.006.
[52] P.P. Das, M. Sharma, M.K. Purkait, Ammonia oxidizing bacteria in wastewater treatment, in: M.P. Shah (Ed.), *Ammonia Oxidizing Bacteria: Applications in Industrial Wastewater Treatment*, Royal Society of Chemistry, 2023: pp. 83–102. https://doi.org/10.1039/BK9781837671960-00083.
[53] M.A. Madhavan, S.P. Antony, Effect of polarity shift on the performance of electrocoagulation process for the treatment of produced water, *Chemosphere* 263 (2021) 128052. https://doi.org/10.1016/j.chemosphere.2020.128052.

[54] P.P. Das, A.D. Sontakke, N.S. Samanta, M.K. Purkait, Emerging contaminants in wastewater: Eco-toxicity and sustainability assessment, in: M.P. Shah (Ed.), *Industrial Wastewater Reuse: Applications, Prospects and Challenges*, Springer Nature Singapore, Singapore, 2023: pp. 63–87. https://doi.org/10.1007/978-981-99-2489-9_4.

[55] S.R.S. Bandaru, A. Roy, A.J. Gadgil, C.M. van Genuchten, Long-term electrode behavior during treatment of arsenic contaminated groundwater by a pilot-scale iron electrocoagulation system, *Water Res.* 175 (2020) 115668. https://doi.org/10.1016/j.watres.2020.115668.

[56] S.E. Amrose, S.R.S. Bandaru, C. Delaire, C.M. van Genuchten, A. Dutta, A. DebSarkar, C. Orr, J. Roy, A. Das, A.J. Gadgil, Electro-chemical arsenic remediation: Field trials in West Bengal, *Sci. Total Environ.* 488–489 (2014) 539–546. https://doi.org/10.1016/j.scitotenv.2013.11.074.

[57] M.H. Ibrahim, D.T. Moussa, M.H. El-Naas, M.S. Nasser, A perforated electrode design for passivation reduction during the electrochemical treatment of produced water, *J. Water Process Eng.* 33 (2020) 101091. https://doi.org/10.1016/j.jwpe.2019.101091.

[58] Y. Yu, Y. Zhong, M. Wang, Z. Guo, Electrochemical behavior of aluminium anode in super-gravity field and its application in copper removal from wastewater by electrocoagulation, *Chemosphere* 272 (2021). https://doi.org/10.1016/j.chemosphere.2021.129614.

[59] C.C. He, C.Y. Hu, S.L. Lo, Evaluation of sono-electrocoagulation for the removal of Reactive Blue 19 passive film removed by ultrasound, *Sep. Purif. Technol.* 165 (2016) 107–113. https://doi.org/10.1016/j.seppur.2016.03.047.

[60] W. Dong, X. Gu, Y. Shu, D. Cao, J. Yu, M.A. Abdel-Fatah, H. Fu, Pulse electrocoagulation combined with a coagulant to remove antimony in wastewater, *J. Water Process Eng.* 47 (2022) 102749. https://doi.org/10.1016/j.jwpe.2022.102749.

[61] N.S. Samanta, P.P. Das, M. Sharma, M.K. Purkait, Recycle of water treatment plant sludge and its utilization for wastewater treatment, in: M. Sillanpää, A. Khadir, K. Gurung (Eds.), *Resource Recovery in Drinking Water Treatment*, Elsevier, 2023: pp. 239–264. https://doi.org/10.1016/B978-0-323-99344-9.00010-4.

4 Mathematical Modeling and Simulation of Electrochemical Reactors

4.1 INTRODUCTION

It is important to note that a number of studies emphasize how parallel plates can be used as electrodes flooded in bakers to characterize various electrochemical processes, with a magnetic stirrer being responsible for the fluid motion [1, 2]. These cells are ideal for exploring ideas of electrochemical processes at a lab scale (less than 500 mL). Low current efficiency is the result of current distribution issues caused by the electrode accommodation in such cells, such as border effects at the electrode corners. Moreover, because of the rotational fluxes and vortex that form inside the baker, hydrodynamics in such cells is difficult to precisely regulate [3]. It is not possible to scale up the current density, electrode potential, and hydrodynamics of these cells in a way that is satisfactory. The electrochemical reactors (ECRs) can be divided into three types according to their mode of operation: batch, continuous, and continuous with recirculation. Another ECR is the cylindrical reactor with swirling flow, in which the outer electrode is placed behind the inner electrode, which can be either conical or cylindrical. The inner electrode of a reactor with a rotating cylinder electrode (RCE) geometry rotates to cause the solution to move. The ECRs with spinning cylinder electrodes and parallel plate electrodes are the ones that typically operate in batch mode [4]. It is possible to characterize the ECR experimentally; however, doing so would cost too much and take up numerous working hours. Nonetheless, the design, characterization, and scaling-up of ECRs have been aided by the recent advancements in CFD approaches. For both basic research and practical uses, a number of scientists have modeled the hydrodynamics, mass and heat transport, potential, and current distributions in numerous ECRs using computational fluid dynamics (CFD) techniques. If a suitable sensibility analysis is carried out, the outcomes of both numerical approaches are comparable [5]. It should entail determining the ideal grid size that will solve a transport attribute or parameter that is size invariant. Numerous CFD simulations of ECRs have been performed for a variety of electrolytic processes, including fuel cells, redox flow batteries, metal-air batteries, chlorination, chlor-alkali, inorganic electrosynthesis, organic electrosynthesis, electrowinning of metals, electroplating, electroforming, electrodialysis (ED), water purification, and environmental treatment. Water treatment via electrocoagulation,

DOI: 10.1201/9781003499671-4

metal ion recovery, and various electrochemical advanced oxidation processes, including anodic oxidation, ozonation, photoelectro-Fenton, and hybrid processes, are the primary environmental applications [6–10]. Before industrial development, CFD can be used to address the range of kinetic charge transfer rates provided by different electrode materials. Everything mentioned above has an impact on the conversion rate, product quality, energy usage, electrolysis energy consumption, and process costs. Given the needs of the modern world, designing ECRs presents significant obstacles. Future advancements, however, are conceivable and include the creation of unique electrode materials and forms, innovative electrochemical processes, sustainable processing in conjunction with current methods, and a reevaluation of established techniques [11].

The significance of simulation and modeling in the fabrication of ECRs utilized in various electrochemical processes is covered in this chapter. Using examples, the enhancement of current reactors after CFD simulations is investigated. The simulation-derived designs of smart ECRs are also discussed. Lastly, the difficulties and future directions for ECR design and the creation of new electrochemical processes are discussed.

4.2 ELECTROCHEMICAL REACTOR DESIGN

In order to assess the behavior of various electrolyzer types, make modifications, and develop new ECR geometries for various electrochemical processes, researchers recently examined them using mathematical models and numerical simulations. Filter-press, conical electrode, rotating disc, rotating ring, rotating cylinder, tubular, stirred tank, and proton-exchange membrane reactors are a few of these electrochemical cell designs [12]. A few of the recently simulated ECRs are shown in Figure 4.1, which the authors direct the reader to. The FM01-LC, a scaled-down version of the FM21-SP (2100 cm^2 area), is a laboratory filter-press type reactor with an electroactive area of 64 cm^2. It is commonly used in a number of electrochemical applications, mainly in the chlor-alkali process, as shown in Figure 4.1(a). Another filter-press type reactor, depicted in Figure 4.1(b), is composed of eight empty channels arranged in a serpentine pattern with a surface area of 144 cm^2. It operates in horizontal mode and is exposed to the atmosphere. Numerous electrochemical water treatment techniques, including electrocoagulation, make use of this reactor [4]. Figure 4.1(c) depicts an electrochemical cell known as Diacell, which has a reaction volume of 117 cm^3, is made up of two circular electrodes (10 cm in diameter), and has a surface area of 70 cm^2 and two spacers. Figure 4.1(d) shows a rotating ring electrode (RRE) reactor. It consists of a cylindrical stainless steel container with an operational volume of 0.016 m^3, a concave bottom, and an internal diameter of 27 cm. The RRE has seven anode rings and seven alternate cathodes (0.14 m ring diameter) spaced 1 cm apart. It is rotated by a variable speed unit. A rotating cylinder electrode (RCE) reactor (119 cm^2), mostly utilized for metal recovery by electrodeposition at the laboratory level, is depicted in Figure 4.1(e) [13, 14]. An example of an interdigitated circular-flat proton-exchange membrane fuel cell (PEMFC) with 0.5 mm-high and 1 mm-wide channels is shown in Figure 4.1(f). The inlet manifold measures 5 mm in width and 1 mm in height. The diameter of the circular gas diffusion layer (GDL) is 9.4 cm;

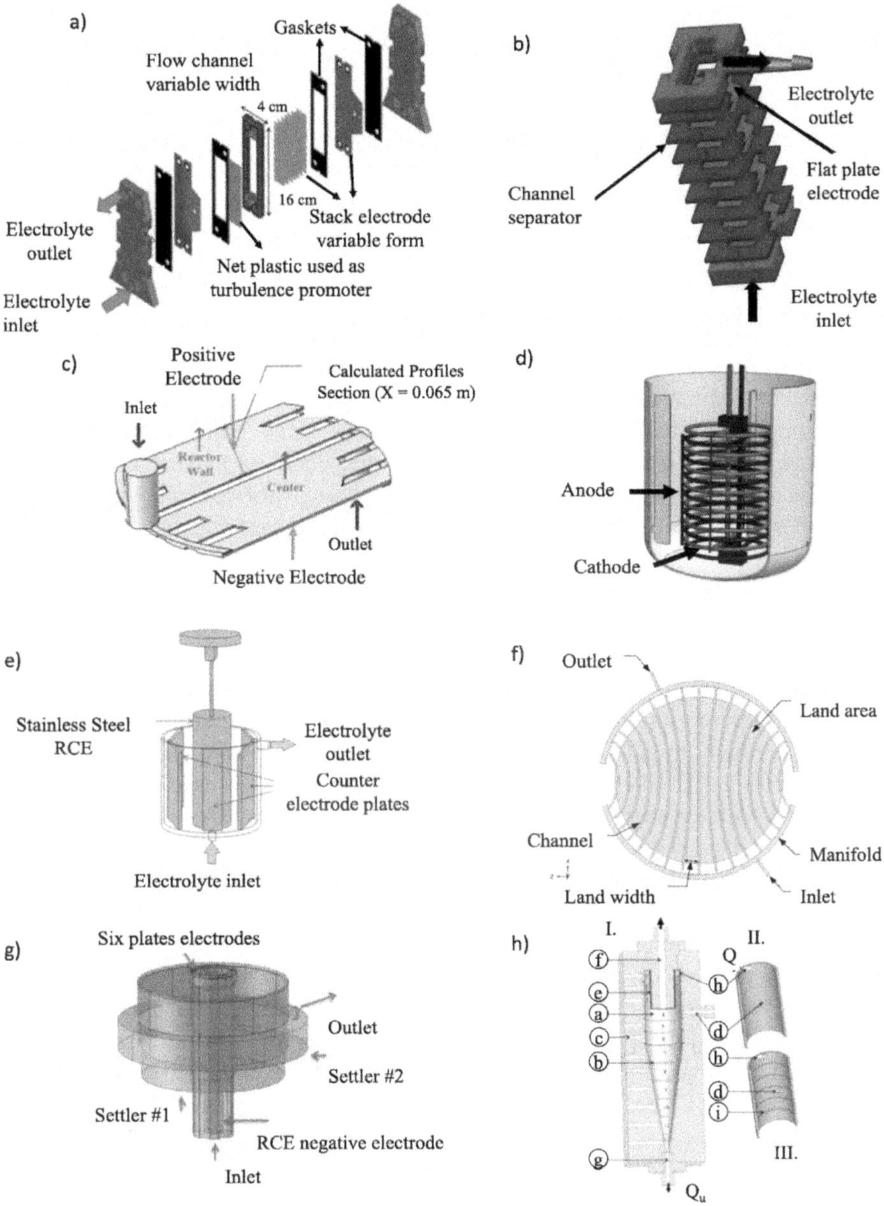

FIGURE 4.1 Several electrochemical cell designs have been investigated through numerical simulations and mathematical models: (a) the filter-press reactor FM01-LC, (b) filter-press reactor in horizontal connection, (c) the electrochemical cell DiaClean, (d) rotating ring electrode reactor (RRE), (e) rotating cylinder electrode (RCE), (f) PEM cell, (g) electrocoagulation reactor design, and (h) hydrocyclone reactor (Reproduced with permission from Rivera et al. (2021) @ Elsevier [6]).

further information is available elsewhere. An electrocoagulation reactor with an RCE in the center connected to two sedimentation tanks is depicted in Figure 4.1(g) [15]. The RCE in this example contains six 30.5 cm-high anodes positioned on the reactor surface at a 60° angle and spaced equally apart from one another. The reactor's bottom is where the input is, and the second settler is where the outlet is. A hydrocyclone utilized as an ECR is shown in Figure 4.1(h). It has three distinct cathodic arrangements (shown as I, II, and III) to process two-phase system (liquid-gas) [16].

4.3 MATHEMATICAL MODELING AND SIMULATION OF THE REACTION ENVIRONMENT

Mathematical models are often used that make use of parametric techniques. When mesh methods cannot be applied to complex geometries, such as packed bed electrode, complex electrode forms, or multi-electrode stack modeling, these classical approaches are frequently used. The one-dimensional parametric equations are solved. Examples from various research groups studying various electrochemical systems and/or applications are used to study the aforementioned mathematical models.

4.3.1 Single-Phase Flow

There are primarily five reactor topologies that have been the subject of several papers, most notably those on rotatory and parallel plate reactors. Also, as the majority of ECRs operate in turbulent environments, the κ-ε and renormalization Group (RNG) κ-ε, which are CFD techniques, are the best choices for assessing the fluid regime inside the ECRs. Its formulation, which incorporates additional tensorial components with anisotropic coefficients, enables a rigorous physical description of turbulence events [17]. Despite these advantages, the κ-ω model is not widely used due to its complexity and the need for hard-to-understand complex coefficients and parameters, making its maintenance difficult. Therefore, the standard κ-ε and its variations are preferred since they have adequately anticipated the turbulence effects on mixing reactor zones, even if the simulation times of the two models are similar. It should be emphasized that the most researched issue in single-phase flow modeling is the assessment of the flow regime in parallel plate reactors with various intake and electrode gap configurations [18]. For instance, Vázquez et al. (2010) used the commercial CFD code Fluent™ to examine the design of a filter-press reactor [19]. Their results show that the low input velocities and the design of the reactor entrances both have an impact on the reactor's flow streamlines.

Furthermore, Litrico et al. (2017) created a model to study the complex boundary conditions and strong variable coupling in equations governing the mass transfer, potential distribution, and flow field in a parallel plate ECR during copper electro refining [20]. They did this using the open-source program Open Foam. To comprehend the function of various arrangements, specialized research has been done. In a tubular electrochemical reactor, for instance, Su et al. (2011) investigated how the arrangement of the electrodes affected the mass transfer rate [21]. Velocity profiles were acquired and electrodes made of plates and mesh were assessed. Based on simulations conducted on various configurations, it has been shown that the increased mass transfer factors in the mesh configuration are a result of the solution being induced to rotate due to high shear rates created in the mesh surface. The impacts of the distributor, also known as the diversion trench, and the flow spacers' length-to-width ratio on the flow characteristics of the fluid were demonstrated by Yuxiang et al. (2018) [22]. The findings there indicated, as reported in other specialist references, that the width and arrangement of the flow distributors have a significant impact on the distribution of flow within this type of reactor. Furthermore, an increase in pressure drop was seen in conjunction with the flow fluid improvement. A number of studies on hydrodynamic modeling for multi-electrode stacks, serpentine array channels, tubular and spinning electrodes, and 3D porous electrodes have recently been published. It is important to note that the NS and RANS techniques are not directly applicable to this type of matrix in the hydrodynamic description of porous electrodes. These methods, referred to as Darcy equations, entail a number of considerations, which are covered in detail in this book. The CFD method has been used to examine these types of reactors for different purposes, such as the electrocoagulation process.

4.3.2 Two-Phase Flow

The modeling of galvanic fuel cells was one of the earliest applications of gas phase dynamics research; nevertheless, there are more complex scenarios in which gaseous phase production occurs at the electrode, necessitating the application of different methodologies. The existence of gas bubbles alters the flow pattern of the dispersed phase, which is the continuous phase (the liquid phase) [23]. Since there are several approaches with varying degrees of rigorousness available, describing a fluid pattern with two phases is typically a complex undertaking. The nomenclature section describes these models' mathematical symbols. In ECR simulation, three primary two-phase model techniques have been utilized: Euler–Euler, Euler–Lagrange, and mixture models [6]. Specific boundary requirements and mathematical closures apply to each of these models. Determining the interaction between the dispersed phase particles, particularly the gas bubbles, and the continuous phase is a challenging task when describing fluid patterns under two-phase settings. Several forces have been proposed to exist for continuous and dispersed interactions, including buoyancy, drag, pressure gradient, additional mass, and lift force [24].

A gas-evolving CFD modeling was carried out by Liu et al. (2018) [25]. Although the liquid velocity motion was precisely estimated by both of the models provided in this work, it was found that the 2D model was unable to account for the fluctuations in velocity that the particle image velocimetry (PIV) system monitored. There was no discussion of a design enhancement idea or a methodical description of how operating conditions affect the outcome. Some recommendations for enhancing the design of this type of continuous reactor are also included in this book. An expanding subject of study is the employment of alternative Eulerian model approaches to reduce processing costs; nevertheless, in many cases, these models necessitated the application of intricate and contentious boundary conditions in order to produce a realistic two-phase flow field description. A popular Eulerian method for performing flow field calculations that take the two-phase scenario into account is the mixture model. For more intricate and specialized applied systems, such as liquid-gas and solid-liquid flow, this model has been used well. Mandin et al. (2008) employed this method, accounting for the bubble dispersion [26]. The impact of electrogenerated bubbles on current density modification was also investigated in order to deduce the modification of transport parameters. It was shown that in the stagnant electrolyte situation, the motion profile of the bubbles had a significant impact on the current density distribution over the electrodes.

4.3.3 Fluid Pressure Drop

To control the pumping requirements, the pressure drop in ECRs must be determined. This is also one of the methods for validating and characterizing CFD calculations in continuous ECRs. Jyothi and Jayanti's (2014) work is considered one among the most significant ones in CFD computations [27]. They implemented rib-based designs, also known as interdigitated and serpentine flow distributor arrays, in RFB energy storage devices. To determine the pressure drop and overall performance, they compared the two geometries. Research indicates that the pressure drop throughout the cell is significantly influenced by the flow distributor geometry. Although these flow distributor designs have been widely employed, it is important to note that they have high pressure drops, ranging from 5 to 100 kPa, which means that a significant amount of energy is required to pump the electrolyte. Since gases are a necessary component of PEMFC applications, flow field designs for proton-exchange membrane fuel cells (PEMFCs) have typically been applied to redox flow batteries (RFBs) without much thought. The performance of various ECR applications and the significance of felt and porous electrodes in mass transport restrictions have been emphasized in recent works, mostly due to their large specific surface area. However, due to the pumping requirements, these electrodes also come at a higher cost [28].

Rivera et al. (2017) showcased that the parallel plate channel requires less pumping energy than the turbulence promoter to solve the Reynolds average Navier-Stokes (RANS) equation; however, these pressure losses were negligible when compared to other rib-based geometries [29]. Yuxiang et al. (2018) used CFD to assess the pressure drop in a tubular reactor channel [22]. They come to the conclusion that the length of the flow distributor has little bearing on the total pressure drop throughout the reactor. According to CFD studies, the small dead zones that result from a reactor's inlet-outlet geometry expanding could be a factor in the pressure decrease in certain systems. The recorded pressure drop in this case was between 550 and 800 Pa. The Brinkmann approach overestimated the experimental results by demonstrating a proportionate link between pressure drop and porosity. It might be explained by the fact that such a model does not incorporate inertial effects. Despite the characteristic non-uniform flow behavior of structured electrodes, the flow patterns created by their employment enable a maximal pressure drop of approximately 400 Pa as opposed to 1000 Pa in the spacer promoter setup.

4.3.4 Intensification of Electrochemical Reactors

Understanding the phenomenological behavior of ECRs through modeling and simulation enables us to make changes to the shape and operating conditions that enhance process performance. The term "process intensification" refers to the latter and has been applied to ECRs. Research projects are currently using modeling and simulation to develop novel reactor designs or to enhance existing ones that are used in various electrochemical processes. In a groundbreaking study, Frías-Ferrer et al. (2011) employed CFD simulations to determine the flow-active and stationary regions in a bench-scale filter-press electrochemical reactor (Figure 4.2 I a), which is primarily utilized for wastewater treatment [30]. These authors used experimental RTD curves to validate the flow calculations. To reduce the sluggish and channeling flow zones, they made modifications to the reactor's inlet and outflow portions. UA63.04 was the name of that reactor (Figure 4.2 I b). Cruz-Díaz et al. (2014) devised a new fabrication for the input and output distributors of the FM01-LC system (Figure 4.2 II a),

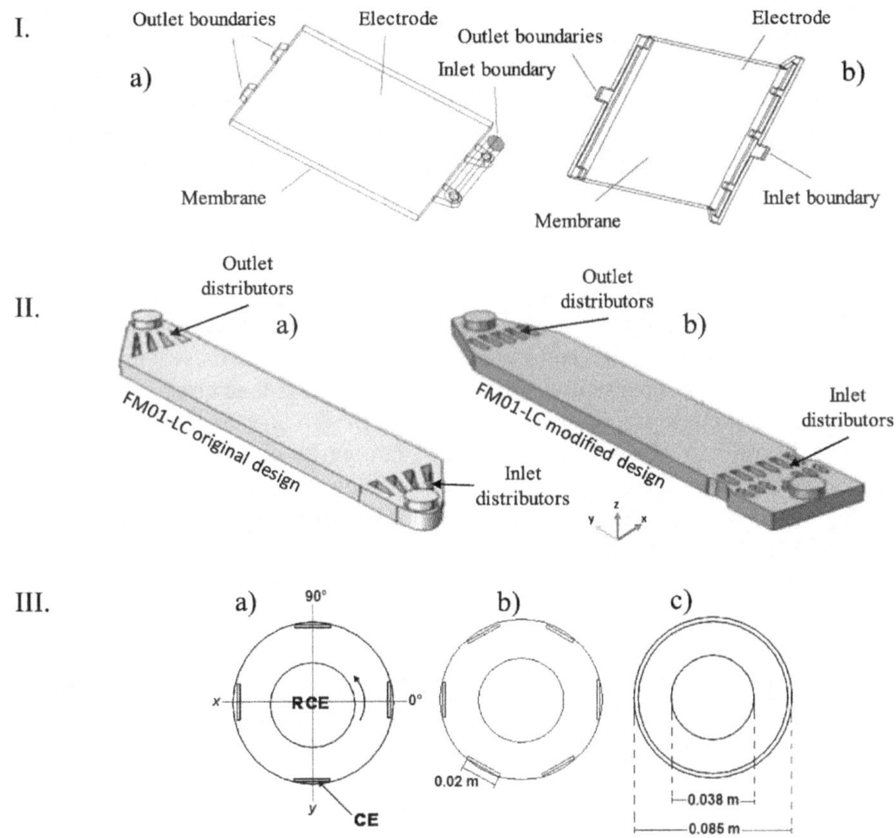

FIGURE 4.2 Different electrochemical reactors enhanced via simulation: I. UA63.03 reactor: (a) real design and (b) enhanced design; II. FM01-LC reactor: (a) real design and (b) enhanced design; and III. Rotating cylinder electrodes: (a) with four-plate counter electrode, (b) with six-plate counter electrode, and (c) concentric cylinders as counter electrodes (Reproduced with permission from Rivera et al. (2021) @ Elsevier [6]).

which is another example of ECR intensification [12]. They used the Reynolds-averaged Navier–Stokes for low Reynolds number (RANS-LRN) equations to solve the hydrodynamics of the redesigned reactor (Figure 4.2 II b). They used the averaged convection-diffusion equation while taking into account a turbulent flow environment to simulate the tracer. Because of the practically constructed electrolyte flow distributors, they were able to achieve a more uniform velocity field across the electroactive zone. Pérez and Nava (2013) created a unique arrangement of rotating cylinder electrode reactors, combining four and six plates with a concentric cylinder as the counter electrode (Figure 4.2 III a, b, and c) [31]. While the hydrodynamics was more uniform and controlled in the concentric design, these vortices enhanced mass transport. There have also been other research publications reporting on the use of this spinning cylinder electrode reactor for various purposes. CFD simulations provide the chance to enhance well-established ECRs for application in a variety of electrochemical processes.

While certain research has been conducted to enhance the reactor geometry and operating parameters through modeling and simulation, the scope of these studies has been restricted to specific cell types. As a result, this field offers the chance to produce fresh research on different kinds of cells. For example, modeling of innovative and feasible energy techniques such as bioelectrochemical reactors and microreactors has been the subject of more research articles recently. In applications such as chemical analysis, sensors, lab-on-a-chip, microflow fuel cells, and others, it facilitates the generation or detection of specific chemicals and allows the management of a given process through the use of a controllable flow regime [32]. Typically, two-inlet designs in electrochemical microreactors are employed to take advantage of the low Reynolds number laminar flow in a dual microchamber in order to produce co-laminarity, also referred to as co-laminar flow. Under these conditions, the reactants can only be separated via diffusive transport at this interface; a membrane is not necessary. High precision is needed in the design and operation of this kind of reactor in order to take advantage of the co-laminarity effect. Microfluidic ECRs are fascinating due to their many possible applications; yet, at the moment, most research on ECR models and simulations is done at the macro-level. It is clear that micro-ECR CFD simulation is a choice that enables the planning and forecasting of their appropriate operation prior to construction [33]. Numerous studies have been conducted on the modeling and simulation of micro-ECRs, primarily with regard to their use in redox flow batteries, various fuel cell types, hydrogen production, and the electrosynthesis of chemicals and pharmaceuticals.

4.4 NOVEL DESIGNS OF ELECTROCHEMICAL REACTORS

An effective method for better comprehending the intricate phenomenology found in an ECR's reaction environment is CFD modeling. To recover metal ions from diluted solutions, Pérez et al. (2020) created a smart filter-press type cell with parallel plate electrodes [34]. They distributed the electrolyte at the entrance using fine fins, and then before the interelectrode area, they created a quiet zone and a net-like spacer to let the electrolyte flow. The fins and plastic net were at the exit, allowing the electrolyte to flow toward it. Additionally, a few parts were created via 3D printing.

Escudero-González and López-Jiménez (2014) evaluated the hydrodynamics and pressure drop using plate electrodes that accounted for 930 cm^2 of the estimated membrane area in a filter-press electrolyzer [35]. This work's ECR was designed to be utilized as an RFB. For homogenization of the flow pattern, complex inlet-outlet manifolds were attached to the reactor. In addition, a rib design was fastened to the inlet-outlet distributors to provide the membrane with mechanical support. RTD and pressure drop measurements were used to assess and validate hydrodynamics. There are numerous types of reactor components in the subject of rotating electrode reactor. For instance, Yáñez-Varela et al. (2020) modified a tubular electrode baffled reactor (EBR) to enable its use in the treatment of Cr(VI) from contaminated effluents [13]. This reactor's performance was contrasted with that of the rotating ring configuration (RRE) reactor, which this research team uses frequently. In summary, this section discusses the role that CFD analysis plays in the development of innovative electrochemical reactors (ECRs) and their applications in environmental applications, energy storage, and conversion.

4.5 PERFORMANCE OF CHEMICAL AND ELECTROCHEMICAL REACTIONS

Establishing the phenomenological and engineering foundation for describing and scaling-up a number of electrochemical and chemical processes at the laboratory, pre-pilot, and industrial scales is another goal of CFD analysis in ECRs.

4.5.1 Flow-Based Electrochemical Energy Storage

Wang et al. (2017) used computational fluid dynamics (CFD) and experimental methods to investigate the impact of the porosity and compression ratio of the gas diffusion layer on a fuel cell system [36]. Enhanced compression ratio in the gas diffusion layer caused the component's porosity to diminish, which helped remove heat from the cell but hampered oxygen diffusion and negatively impacted the overall current density and cathode activation loss. A study conducted by Wilberforce et al. (2019) assessed the effects of several membrane-electrode assemblies (MEAs) or flow channels in a traditional PEMFC device [37]. They investigated unique flow arrangements and contrasted them with the standard serpentine configuration design. When cellular porous aluminum foam was used in the electrode and flow assemblies' simulation, the pressure drop was significantly reduced, reaching just 30 or 40 Pa. The device was optimized using numerical data, and new flow channels and electrode assembly configurations were physically manufactured. The unique device outperformed the standard serpentine device, as demonstrated by the comparison between the two. In a similar vein, Atyabi and Afshari (2019) used a PEMFC simulation to study the flow field of a sinusoidal geometry [38]. The velocity distributions for the straight and sinusoidal geometry are presented in Figure 4.3a. The sinusoidal channel yields the maximum current, which is 1.8 times higher than the straight configuration's, as shown in Figure 4.3b. Due to the development of Dean vortices at sinusoidal bend geometries, this impact may boost mass diffusion.

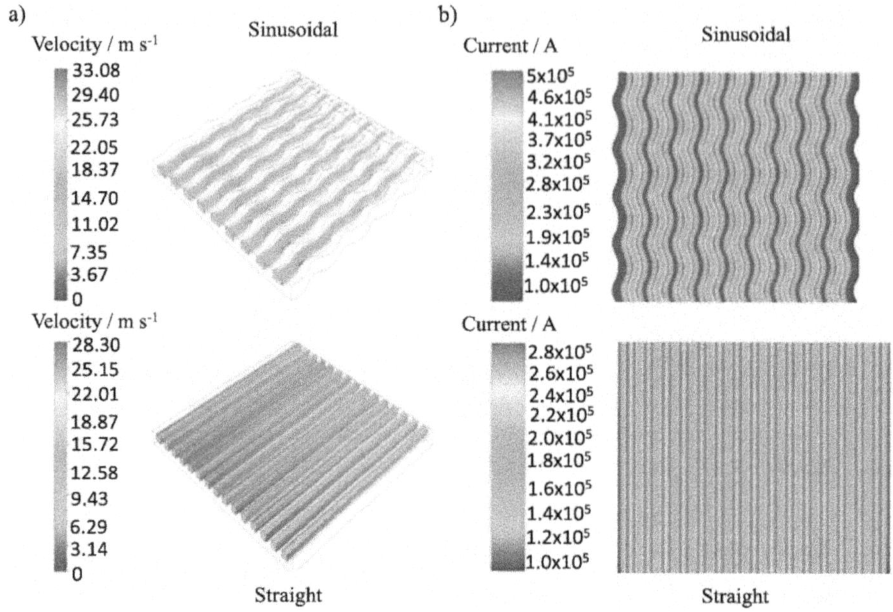

FIGURE 4.3 A PEMFC fuel cell system using CFD modeling compares the flow channel geometries of sinusoidal and straight streams: (a) velocity distribution contour and (b) local current distribution (Reproduced with permission from Rivera et al. (2021) @ Elsevier [6]).

In their comprehensive description of an all-vanadium redox flow battery (VRFB), Gandomi et al. (2016) employed the Darcy technique to compute the hydrodynamics in a porous electrode [39]. Taking into account the potential decreases in the electrode, electrolyte, and membrane sections, a 1D model was used in this work to numerically solve the distribution of chemical species. During the cycle tests, numerical data were compared with experimental potential measurements made using individual reference probes in various battery parts. These findings demonstrated that as the flow rate increased during the discharge, the current density also increased, suggesting that such RFB worked under conditions that limited mass movement.

4.5.2 Electrosynthesis and Environmental Applications

Environmental electrochemical applications have received a lot of attention, according to several reputable reviews. Several issues have been brought forward, including those related to metal recovery, organic pollution remediation using powerful oxidant electrogeneration, bioelectrochemical systems, and electrocoagulation. A number of engineering issues are raised by all these communications, most of which deal with sophisticated mechanisms in the bulk region in addition to electrochemical reactions [40]. It serves as an illustration of the intricate relationship that exists between operational conditions and ECR performance. Comprehending these intricate mechanisms in the bulk region will be crucial for managing the efficacy of ECRs utilized on medium and big scales. Through the use of CFD research, environmental and engineering applications are connected, enabling a deeper understanding of the intricate reaction mechanisms and operational parameters of these systems. As a result, several attempts have been made to combine distinct electrode and reaction kinetics in electrocoagulation, strong oxidant production, metal recovery, and the removal of developing contaminants. These investigations often look for calculations of the performance of the specific reactor and how it relates to standard operating factors [41].

Using electrocoagulation and mediated oxidation, Körbahti (2014) modeled a tubular electrochemical reactor for wastewater remediation [42]. Two weighted response terms and a convection-diffusion connection made up the governing equations in this case. After studying the relationship between geometry and the residence time effect, it was found that, depending on the kind of wastewater used, reduction of the residence time resulted in the increase of chemical oxygen demand (COD) removal. The impact of hydrodynamics within a filter-press reactor on copper electrodeposition was systematically studied by Rivero et al. (2015) [43]. The results and experimental validation point to an increase in copper removal with increasing flow rate, indicating that mass transport is normally responsible for controlling and limiting this metal's removal. The active chlorine process has been the subject of numerous mathematical investigations; however, most of these have published only parametric model usage. As stated in, there are presently few CFD investigations on the production of chlorine species. To explain the experimental data, Palma-Goyes et al. (2019) offered a CFD approach; on Sb_2O_3-doped Ti/RuO_2-ZrO_2 anodes, active chlorine was generated [44]. The cathodic reaction in the undivided cell was a part of the simulated mechanism, and there was a very good agreement between the simulations and the experiments. Pérez et al. (2017) provided another example of mathematical modeling by simulating a solar photoelectro-Fenton flow plant during the breakdown of erythromycin [45]. Because the operation of the flow reactor plant is so complex, a parametric model was used to carry out the intricate electrochemical advanced oxidation process. The impact of the initial pharmacological concentration, current density, and mean linear flow rate on the breakdown of total organic carbon was discussed. The results obtained from the experiment and theory were in good agreement owing to the total organic carbon monitoring for electrolysis.

4.5.3 Scaling-Up Cases

CFD simulations and mathematical modeling are effective methods for increasing ECRs in size. There are few publications that demonstrate scaling-up techniques. For example, Almazán-Ruiz et al. (2012) [46] conducted a study to scale up a rotating cylinder electrode using computational fluid dynamics (CFD), while Wang et al. (2015) [47] did the same with an electrochemical tubular reactor. A divided filter-press type electrolyzer was scaled up by Nava et al. (2008) for the dual purposes of cathodic recovery of gold and silver and the simultaneous synthesis of formamidine disulfide, a selective oxidant that leaches gold and silver from minerals [48]. Owing to the electrolyzer's and the process's overall complexity, a parametric model incorporating mass transport, hydrodynamics, and potential and current distributions was created; simulations and experiments showed good agreement in this model. One possible explanation for the dearth of scale-up studies is the confidentiality surrounding industrial development or intellectual property protection.

4.6 CONCLUSION

It has been shown that using mathematical techniques to create new ECRs for both basic research and practical uses can be beneficial when creating certain cell components, including electrode sizes and shapes. Through the modeling of fluid dispersion, mass transport, and potential and current distributions, they have also been proved helpful in describing the reaction environment. These kinds of theoretical investigations have been conducted using 1D parametric models and 3D CFD simulations. The proton-exchange membrane fuel cell, which has undergone extensive optimization through CFD simulations, is the most sophisticated technology made possible by these computational methods. The intensification of some current ECRs that operate in unfavorable settings in a number of classical processes has been made possible owing in large part to CFD models. In conclusion, commercial applications have proved the viability of the methodology based on the use of CFD to design technologies, as is the case with redox flow batteries and fuel cells. The use of CFD tools has occasionally been undervalued despite its enormous potential to improve a number of electrochemical processes. Despite this, as this review has demonstrated, a thorough analysis of ECRs using CFD tools necessitates the application of numerous scientific ideas, including a thorough grasp of electrochemistry, as well as experience formulating problems in ECRs from the basic transport equation and overcoming a substantial number of technical and research challenges.

REFERENCES

[1] P.P. Das, A.D. Sontakke, M.K. Purkait, Electrocoagulation process for wastewater treatment: Applications, challenges, and prospects, in: M.P. Shah (Ed.), *Development in Wastewater Treatment Research and Processes*, Elsevier, 2023: pp. 23–48. https://doi.org/10.1016/B978-0-323-95684-0.00015-4.

[2] P.P. Das, M. Sharma, M.K. Purkait, Recent progress on electrocoagulation process for wastewater treatment: A review, *Sep. Purif. Technol.* 292 (2022) 121058. https://doi.org/10.1016/j.seppur.2022.121058.

[3] S. Li, B. Sundén, Effects of gas diffusion layer deformation on the transport phenomena and performance of PEM fuel cells with interdigitated flow fields, *Int. J. Hydrogen Energy* 43 (2018) 16279–16292. https://doi.org/10.1016/j.ijhydene.2018.07.064.

[4] L.F. Castañeda, J.L. Nava, Simulations of single-phase flow in an up-flow electrochemical reactor with parallel plate electrodes in a serpentine array, *J. Electroanal. Chem.* 832 (2019) 31–39. https://doi.org/10.1016/j.jelechem.2018.10.030.

[5] F.F. Rivera, C.P. de León, J.L. Nava, F.C. Walsh, The filter-press FM01-LC laboratory flow reactor and its applications, *Electrochim. Acta* 163 (2015) 338–354. https://doi.org/10.1016/j.electacta.2015.02.179.

[6] F.F. Rivera, T. Pérez, L.F. Castañeda, J.L. Nava, Mathematical modeling and simulation of electrochemical reactors: A critical review, *Chem. Eng. Sci.* 239 (2021) 116622. https://doi.org/10.1016/j.ces.2021.116622.

[7] P.P. Das, S. Dhara, M.K. Purkait, Hybrid electrocoagulation and ozonation techniques for industrial wastewater treatment, in: M.P. Shah (Ed.), *Sustainable Industrial Wastewater Treatment and Pollution Control*, Springer Nature Singapore, Singapore, 2023: pp. 107–128. https://doi.org/10.1007/978-981-99-2560-5_6.
[8] P.P. Das, P. Mondal, Anweshan, A. Sinha, P. Biswas, S. Sarkar, M.K. Purkait, Treatment of steel plant generated biological oxidation treated (BOT) wastewater by hybrid process, *Sep. Purif. Technol.* 258 (2021) 118013. https://doi.org/10.1016/j.seppur.2020.118013.
[9] P.P. Das, A. Anweshan, P. Mondal, A. Sinha, P. Biswas, S. Sarkar, M.K. Purkait, Integrated ozonation assisted electrocoagulation process for the removal of cyanide from steel industry wastewater, *Chemosphere* 263 (2021) 128370. https://doi.org/10.1016/j.chemosphere.2020.128370.
[10] P.P. Das, A. Anweshan, M.K. Purkait, Treatment of cold rolling mill (CRM) effluent of steel industry, *Sep. Purif. Technol.* 274 (2021). https://doi.org/10.1016/j.seppur.2021.119083.
[11] L.F. Catañeda, F.F. Rivera, T. Pérez, J.L. Nava, Mathematical modeling and simulation of the reaction environment in electrochemical reactors, *Curr. Opin. Electrochem.* 16 (2019) 75–82. https://doi.org/10.1016/j.coelec.2019.04.025.
[12] M.R. Cruz-Díaz, E.P. Rivero, F.J. Almazán-Ruiz, Á. Torres-Mendoza, I. González, Design of a new FM01-LC reactor in parallel plate configuration using numerical simulation and experimental validation with residence time distribution (RTD), *Chem. Eng. Process. Process Intensif.* 85 (2014) 145–154. https://doi.org/10.1016/j.cep.2014.07.010.
[13] J.A. Yáñez-Varela, A. Alonzo-Garcia, I. González-Neria, V. Mendoza-Escamilla, G. Rivadeneyra-Romero, S.A. Martínez-Delgadillo, Experimental and numerical evaluation of the performance of the electrochemical reactor operated with static and dynamic electrodes in the reduction of hexavalent chromium, *Chem. Eng. J.* 390 (2020) 124575. https://doi.org/10.1016/j.cej.2020.124575.
[14] M. Rosales, T. Pérez, J.L. Nava, Computational fluid dynamic simulations of turbulent flow in a rotating cylinder electrode reactor in continuous mode of operation, *Electrochim. Acta* 194 (2016) 338–345. https://doi.org/10.1016/j.electacta.2016.02.076.
[15] A.D. Villalobos-Lara, T. Pérez, A.R. Uribe, J.A. Alfaro-Ayala, J. de J. Ramírez-Minguela, J.I. Minchaca-Mojica, CFD simulation of biphasic flow, mass transport and current distribution in a continuous rotating cylinder electrode reactor for electrocoagulation process, *J. Electroanal. Chem.* 858 (2020) 113807. https://doi.org/10.1016/j.jelechem.2019.113807.
[16] A.N. Colli, J.M. Bisang, Time-dependent mass-transfer behaviour under laminar and turbulent flow conditions in rotating electrodes: A CFD study with analytical and experimental validation, *Int. J. Heat Mass Transf.* 137 (2019) 835–846. https://doi.org/10.1016/j.ijheatmasstransfer.2019.03.152.
[17] A.J. Karabelas, M. Kostoglou, C.P. Koutsou, Modeling of spiral wound membrane desalination modules and plants—review and research priorities, *Desalination* 356 (2015) 165–186. https://doi.org/10.1016/j.desal.2014.10.002.
[18] G. Leonzio, Study of mixing systems and geometric configurations for anaerobic digesters using CFD analysis, *Renew. Energy* 123 (2018) 578–589. https://doi.org/10.1016/j.renene.2018.02.071.
[19] L. Vázquez, A. Alvarez-Gallegos, F.Z. Sierra, C. Ponce de León, F.C. Walsh, Simulation of velocity profiles in a laboratory electrolyser using computational fluid dynamics, *Electrochim. Acta* 55 (2010) 3437–3445. https://doi.org/10.1016/j.electacta.2009.08.066.
[20] G. Litrico, C.B. Vieira, E. Askari, P. Proulx, Strongly coupled model for the prediction of the performances of an electrochemical reactor, *Chem. Eng. Sci.* 170 (2017) 767–776. https://doi.org/10.1016/j.ces.2016.12.060.

[21] J. Su, H.-Y. Lu, H. Xu, J.-R. Sun, J.-L. Han, H.-B. Lin, Mass transfer enhancement for mesh electrode in a tubular electrochemical reactor using experimental and numerical simulation method, *Russ. J. Electrochem.* 47 (2011) 1293–1298. https://doi.org/10.1134/S1023193511110140.
[22] Y. Jia, C. Yan, L. Chen, Y. Hu, Optimized design of spacer in electrodialyzer using CFD simulation method, *J. Ocean Univ. China* 17 (2018) 603–608. https://doi.org/10.1007/s11802-018-3397-x.
[23] P.P. Das, S. Dhara, M.K. Purkait, Ozone-based oxidation processes for the removal of pharmaceutical products from wastewater, in: M.P. Shah, P. Ghosh (Eds.), *Development in Wastewater Treatment Research and Processes*, Elsevier, 2024: pp. 287–308. https://doi.org/10.1016/B978-0-443-19207-4.00003-3.
[24] J. Schillings, O. Doche, J. Deseure, Modeling of electrochemically generated bubbly flow under buoyancy-driven and forced convection, *Int. J. Heat Mass Transf.* 85 (2015) 292–299. https://doi.org/10.1016/j.ijheatmasstransfer.2015.01.121.
[25] L. Liu, W. Cai, Y. Chen, Y. Wang, Fluid dynamics and mass transfer study of electrochemical oxidation by CFD prediction and experimental validation, *Ind. Eng. Chem. Res.* 57 (2018) 6493–6504. https://doi.org/10.1021/acs.iecr.7b04226.
[26] P. Mandin, A.A. Aissa, H. Roustan, J. Hamburger, G. Picard, Two-phase electrolysis process: From the bubble to the electrochemical cell properties, *Chem. Eng. Process. Process Intensif.* 47 (2008) 1926–1932. https://doi.org/10.1016/j.cep.2007.10.018.
[27] T. Jyothi Latha, S. Jayanti, Hydrodynamic analysis of flow fields for redox flow battery applications, *J. Appl. Electrochem.* 44 (2014) 995–1006. https://doi.org/10.1007/s10800-014-0720-0.
[28] R. Gundlapalli, S. Jayanti, Effect of channel dimensions of serpentine flow fields on the performance of a vanadium redox flow battery, *J. Energy Storage* 23 (2019) 148–158. https://doi.org/10.1016/j.est.2019.03.014.
[29] F.F. Rivera, L. Castañeda, P.E. Hidalgo, G. Orozco, Study of hydrodynamics at AsahiTM prototype electrochemical flow reactor, using computational fluid dynamics and experimental characterization techniques, *Electrochim. Acta* 245 (2017) 107–117. https://doi.org/10.1016/j.electacta.2017.05.134.
[30] Á. Frías-Ferrer, I. Tudela, O. Louisnard, V. Sáez, M.D. Esclapez, M.I. Díez-García, P. Bonete, J. González-García, Optimized design of an electrochemical filter-press reactor using CFD methods, *Chem. Eng. J.* 169 (2011) 270–281. https://doi.org/10.1016/j.cej.2011.02.053.
[31] T. Pérez, J.L. Nava, Simulation of turbulent flow of a rotating cylinder electrode. Influence of using plates and concentric cylinder as counter electrodes, *Int. J. Electrochem. Sci.* 8 (2013) 4690–4699. https://doi.org/10.1016/S1452-3981(23)14632-3.
[32] J. Zhang, S. Yan, D. Yuan, G. Alici, N.T. Nguyen, M. Ebrahimi Warkiani, W. Li, Fundamentals and applications of inertial microfluidics: A review, *Lab Chip* 16 (2016) 10–34. https://doi.org/10.1039/c5lc01159k.
[33] J. Marschewski, P. Ruch, N. Ebejer, O. Huerta Kanan, G. Lhermitte, Q. Cabrol, B. Michel, D. Poulikakos, On the mass transfer performance enhancement of membraneless redox flow cells with mixing promoters, *Int. J. Heat Mass Transf.* 106 (2017) 884–894. https://doi.org/10.1016/j.ijheatmasstransfer.2016.10.030.
[34] T. Pérez, L.F. Arenas, D. Villalobos-Lara, N. Zhou, S. Wang, F.C. Walsh, J.L. Nava, C.P. de León, Simulations of fluid flow, mass transport and current distribution in a parallel plate flow cell during nickel electrodeposition, *J. Electroanal. Chem.* 873 (2020) 114359. https://doi.org/10.1016/j.jelechem.2020.114359.
[35] J. Escudero-González, P. Amparo López-Jiménez, Redox cell hydrodynamic modelling: Towards real improved geometry based on CFD analysis, *Eng. Appl. Comput. Fluid Mech.* 8 (2014) 435–446. https://doi.org/10.1080/19942060.2014.11015527.

[36] J. Wang, J. Yuan, B. Sundén, Modeling of inhomogeneous compression effects of porous GDL on transport phenomena and performance in PEM fuel cells, *Int. J. Energy Res.* 41 (2017) 985–1003. https://doi.org/10.1002/er.3687.

[37] T. Wilberforce, F.N. Khatib, O.S. Ijaodola, E. Ogungbemi, Z. El-Hassan, A. Durrant, J. Thompson, A.G. Olabi, Numerical modelling and CFD simulation of a polymer electrolyte membrane (PEM) fuel cell flow channel using an open pore cellular foam material, *Sci. Total Environ.* 678 (2019) 728–740. https://doi.org/10.1016/j.scitotenv.2019.03.430.

[38] S.A. Atyabi, E. Afshari, A numerical multiphase CFD simulation for PEMFC with parallel sinusoidal flow fields, *J. Therm. Anal. Calorim.* 135 (2019) 1823–1833. https://doi.org/10.1007/s10973-018-7270-3.

[39] Y.A. Gandomi, D.S. Aaron, T.A. Zawodzinski, M.M. Mench, In situ potential distribution measurement and validated model for all-vanadium redox flow battery, *J. Electrochem. Soc.* 163 (2016) A5188. https://doi.org/10.1149/2.0211601jes.

[40] C.T.J. Low, C. Ponce de Leon, F.C. Walsh, The rotating cylinder electrode (RCE) and its application to the electrodeposition of metals, *Aust. J. Chem.* 58 (2005) 246–262. https://doi.org/doi.org/10.1071/CH05034.

[41] G. Coria, T. Pérez, I. Sirés, E. Brillas, J.L. Nava, Abatement of the antibiotic levofloxacin in a solar photoelectro-Fenton flow plant: Modeling the dissolved organic carbon concentration-time relationship, *Chemosphere* 198 (2018) 174–181. https://doi.org/10.1016/j.chemosphere.2018.01.112.

[42] B.K. Körbahti, Finite element modeling of continuous flow tubular electrochemical reactor for industrial and domestic wastewater treatment, *J. Electrochem. Soc.* 161 (2014) E3225. https://doi.org/10.1149/2.025408jes.

[43] E.P. Rivero, M.R. Cruz-Díaz, F.J. Almazán-Ruiz, I. González, Modeling the effect of non-ideal flow pattern on tertiary current distribution in a filter-press-type electrochemical reactor for copper recovery, *Chem. Eng. Res. Des.* 100 (2015) 422–433. https://doi.org/10.1016/j.cherd.2015.04.036.

[44] R.E. Palma-Goyes, F.F. Rivera, J. Vazquez-Arenas, Heterogeneous model to distinguish the activity of electrogenerated chlorine species from soluble chlorine in an electrochemical reactor, *Ind. Eng. Chem. Res.* 58 (2019) 22399–22407. https://doi.org/10.1021/acs.iecr.9b05185.

[45] T. Pérez, I. Sirés, E. Brillas, J.L. Nava, Solar photoelectro-Fenton flow plant modeling for the degradation of the antibiotic erythromycin in sulfate medium, *Electrochim. Acta* 228 (2017) 45–56. https://doi.org/10.1016/j.electacta.2017.01.047.

[46] F.J. Almazán-Ruiz, F. V Caballero, M.R. Cruz-Díaz, E.P. Rivero, I. González, Scale-up of rotating cylinder electrode electrochemical reactor for Cu(II) recovery: Experimental and simulation study in turbulence regimen, *Electrochim. Acta* 77 (2012) 262–271. https://doi.org/10.1016/j.electacta.2012.06.003.

[47] J. Wang, T. Li, M. Zhou, X. Li, J. Yu, Characterization of hydrodynamics and mass transfer in two types of tubular electrochemical reactors, *Electrochim. Acta* 173 (2015) 698–704. https://doi.org/10.1016/j.electacta.2015.05.135.

[48] J.L. Nava, M.T. Oropeza, C. Ponce de León, J. González-García, A.J. Frías-Ferrer, Determination of the effective thickness of a porous electrode in a flow-through reactor; effect of the specific surface area of stainless steel fibres, used as a porous cathode, during the deposition of Ag(I) ions, *Hydrometallurgy* 91 (2008) 98–103. https://doi.org/10.1016/j.hydromet.2007.12.001.

5 Application of the Electrocoagulation Process in Drinking Water Treatment

5.1 INTRODUCTION

Growing population, industrialization, and non-sustainable water resources are some of the factors contributing to the global concerns of population and water scarcity. By 2020, just 5.8 billion people or 74% of the world's population had access to "safely managed" drinking water sources [1, 2]. The sorting shows better water sources nearby, easily accessible, and uncontaminated by priority chemicals and feces. Nevertheless, essential services are still inaccessible to 2 billion people, among whom 1.2 billion lack even basic water facilities. Approximately 80% of these individuals live in rural regions, with half of them situated in the most underdeveloped nations [3]. By 2030, the Sustainable Development Goals of the United Nations want to ensure the safe, affordable, and equitable distribution of drinking water. Alongside these objectives, assisting developing countries with water-related projects like desalination, recycling, reuse, wastewater treatment, and efficiency enhancements is a major focus [1]. This has sparked research into creating inexpensive, readily maintained, and eco-friendly technology for recycling and treating water in rural or deprived regions. Accessing safe drinking water these days is hard and scarce; water pollution can harm aquatic ecosystems and freshwater supplies. However, more potent and efficient water treatment methods are needed to address the risks posed by pollutants in the environment, such as high concentrations of pollutants. Electrocoagulation (EC) may eliminate a wide variety of water pollutants, such as organic matter (OM), suspended solids (SS), metals and metalloids such as chromium (Cr), iron (Fe), arsenic (As), fluoride (F), and phosphate (P), and bacteria and viruses. According to Sillanpaa et al. (2018), EC is a technically promising method that is extensively employed for the removal of contaminants from drinking water. It functions as an alternative to chemical coagulation (CC), which comprises the electrodissolution of electrodes to introduce metallic coagulants into water. EC integrates flotation, electrochemistry, and coagulation into a single system to offer several benefits [4, 5]. The mechanism of the EC process is depicted in Figure 5.1, where an external DC power supply is connected to an electrolytic cell that has two electrodes (a cathode and an anode) submerged in a conducting solution that needs to be treated. The EC procedure consists of the following steps:

DOI: 10.1201/9781003499671-5

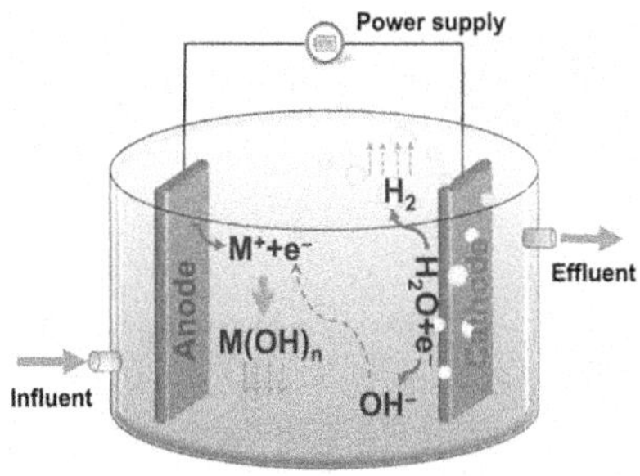

FIGURE 5.1 Schematic diagram of the electrocoagulation process (Reproduced from Mao et al. (2023) @ MDPI [3]).

(1) When an external DC power supply is connected to the EC system, an electric current is passed through it. This process leads to the release of active coagulant cations, which are typically made of aluminum or iron, into the solution. This is done through electrolytic oxidation of a sacrificial anode. Equations (5.1) and (5.4) are used to illustrate this process.
(2) The hydroxyl ions are sequentially produced through cathode hydrolysis, as demonstrated in Equation (5.6).
(3) Monomeric and polymeric species are produced by the interactions of the metal cations with hydroxyl ions as depicted in Equations (5.2), (5.3), and (5.5).
(4) The surface charges of contaminants, suspended particulate matter, and produced emulsions are neutralized when reacted with the metal hydroxyls.
(5) In the aqueous phase, the neutralized particles clump together and coagulate to produce flocs.
(6) The heavier flocs precipitate through sedimentation via sweep coagulation.
(7) The water electrolysis at the cathode produces hydrogen bubbles, which cause flocs to float to the surface of the solution by sweep coagulation, as illustrated in Equation (5.6).

Anode reactions:

$$Fe(s) \rightarrow Fe^{n+}(aq) + ne^{-1} \tag{5.1}$$

$$4Fe^{2+}(aq) + 10H_2O + O_2(aq) \rightarrow 4Fe(OH)_3(s) + 8H^+ \tag{5.2}$$

$$Fe^{2+}(aq) + 2OH^- \rightarrow Fe(OH)_2(s) \tag{5.3}$$

$$Al(s) \rightarrow Al^{3+}(aq) + 3e^- \tag{5.4}$$

$$Al^{3+}(aq) + nH_2O \rightarrow Al(OH)_n^{3-n}(s) + nH^+ \tag{5.5}$$

Cathode reactions:

$$2e^- + 2H_2O \rightarrow H_2 + 2OH^- \tag{5.6}$$

The dissolution of the anode adheres to Faraday's law, as shown in the below equation,

$$m_A = \frac{I * t * M_w}{z * F} \tag{5.7}$$

where M_w signifies the mass of the dissolved anode (g), I indicates the current (A), t is the EC treatment time (s), Mw represents the molecular weight of the anode material (e.g., aluminum with a value of 29.89 g/molmol), F indicates Faraday's constant (with a value of 96,487 C/mol), and z shows the number of electrons involved in the reaction.

Some researchers have proposed that the generation of oxygen at the anode could occur under alkaline pH conditions and when the anodic potential is sufficiently high, as shown in the below equation [6].

$$2H_2O \rightarrow O_2 + 4H^+ + 4e^- \tag{5.8}$$

5.2 CONTAMINATION IN DRINKING WATER AND ITS ASSOCIATED HEALTH IMPACTS

One of the most important requirements for good health is access to an adequate amount of safe drinking water. Nevertheless, waterborne diseases continue to be a leading cause of mortality worldwide, especially among children, and they pose a serious financial burden on many poor countries. There are two primary sources of drinking water: groundwater and surface waterways like rivers and reservoirs, as presented in Table 5.1. All water is contaminated by natural sources, especially inorganic pollutants that come from the geological layers the water runs through. Chemical contaminants in drinking water encompass a wide range of substances, including heavy metals, pesticides, industrial chemicals, and pharmaceuticals. Heavy metals like lead, arsenic, mercury, and cadmium can leach into water from natural deposits, industrial activities, and corroded plumbing systems. Nitrates and nitrites primarily originate from agricultural fertilizers and animal waste. In addition, sediment and turbidity are physical contaminants that can affect water quality and aesthetics. Sedimentation occurs naturally in water bodies, but human activities like construction, deforestation, and erosion can accelerate sediment runoff into rivers and lakes. High turbidity levels not only reduce water clarity but also provide a habitat for bacteria and other microorganisms, necessitating filtration and sedimentation processes during water treatment. Figure 5.2 depicts the various contaminants which are present in drinking water.

5.2.1 Arsenic

Many regions of the world, such as the Indian subcontinent (especially Bangladesh and Bengal), South America, and the Far East, have high rates of sickness due to waterborne arsenic poisoning. It is the only pollutant that causes human cancer when consumed through drinking water. In addition to malignancies of the skin, lungs, bladder,

TABLE 5.1
Contaminants in drinking water and their sources.

Contaminants	Sources	References
Suspended solids	Soil erosion, water runoff, and algal growth	[10]
Dissolved solids (below 2 mm)	Water seeping through soil layers due to both natural and man-made processes	[10]
Heavy metals (such as Ni, Pb, Hg, As, Cu, and Cr)	Industrial waste, corrosion, and erosion	[11]
Fluoride (F)	Naturally exists in groundwater as well as surface water	[12]
Nitrite and nitrate	Utilising preservatives, the nitrogen cycle, fertilisers, and agricultural runoff	[13]
Iron	Both natural sources and pipe corrosion	[11, 12]
Manganese	Naturally occurring in both surface and groundwater, along with iron	[11]
Natural organic matter (NOM)	From plant and animal decomposition	[14]
Polyaromatic hydrocarbons (PAHs)	From utilising fuels based on hydrocarbons	[15]

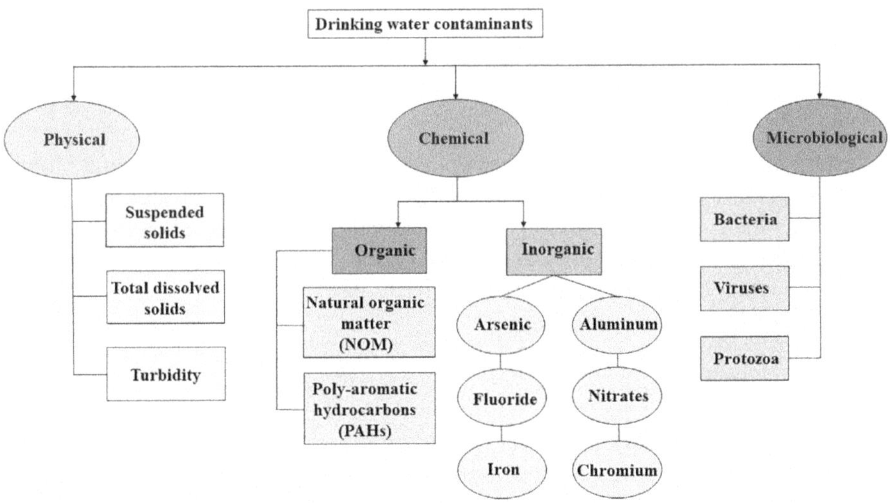

FIGURE 5.2 Contaminants in drinking water.

and possibly the liver, arsenic causes several negative side effects, such as hyperkeratosis and peripheral vascular disease. Nonetheless, the epidemiological data also show the importance of other local parameters, such as nutritional conditions. Determining the amount of arsenic exposure is quite challenging. Millions of tube wells have been sunk in Bangladesh; however, even within short distances, wells can differ greatly in their arsenic levels. Following the achievable limit of achievability, the WHO has set a preliminary guideline value of 10 μg/L. However, there is a continuing discussion regarding the scientific basis for this guideline, including whether the data would allow for a distinction among standard values of, say, 5, 10, or 15 μg/L and whether exposure to 50 μg/L, the previous guideline, would result in illness [7].

5.2.2 Fluoride

In several regions of the world, waterborne fluoride is another significant source of sickness, particularly the Indian subcontinent, Africa, and the Far East, where fluoride levels can surpass 10 mg/L. The fluoride consumption in elevated amounts may cause dental fluorosis, an ugly brown speckling on teeth. Moreover, consuming too much fluoride can lead to skeletal fluorosis, a condition characterized by elevated bone density that may lead to bone fractures and skeletal deformities. The World Health Organization working group determined that a daily total fluoride intake of 14 mg causes skeletal fluorosis and increases the risk of bone fractures. Additionally, there is evidence that fluoride intake higher than around 6 mg may result in an increased risk of bone damage. This is one of the main causes of morbidity and can show symptoms at a young age, making it difficult for affected people to work and possibly making them physically and financially disadvantaged for the rest of their lives. Numerous factors seem to affect the likelihood of negative consequences, such as the amount of water consumed, one's nutritional state, and most importantly the amount of fluoride ingested from other sources [8].

5.2.3 Iron and manganese

In certain anaerobic sources of waterways, significant amounts of iron and manganese have been reported. They become oxidized to low-solubility oxides as the water becomes aerated. Even in quantities much below those that pose a health risk, these will produce noticeable turbidity and discoloration. However, they might force customers to switch to less harmful but potentially more microbiologically dangerous and aesthetically pleasing substitutes [9].

5.2.4 Microplastics

Plastic products are highly useful for human production and existence, but a significant amount of plastic waste is released into the environment. For instance, polymers like polystyrene (PS) and polyethylene (PE) often do not decompose entirely in the natural environment; instead, they progressively shrink and transform into tiny particles known as microplastics (MPs). Numerous studies have demonstrated that MPs can enter the human circulatory system and move up through the food chain, posing a serious threat to human beings. Therefore, governments are acting to limit the amount of plastic items that enter the environment in light of the rising pollution caused by plastic and microplastics (MPs) worldwide.

Drinking water contamination beyond acceptable levels can lead to various health issues based on the type of contaminants, as shown in Table 5.2. The consumption of polluted water is a major contributor to 80% of diseases in developing nations.

TABLE 5.2
Impact of drinking water contaminants on human health.

Contaminants	Health risks associated with prolonged exposure beyond maximum contamination levels	References
Arsenic	A substance known for its carcinogenic properties that may lead to cancers affecting the lungs, bladder, kidneys, and prostate. Moreover, it can lead to skin disorders, vascular problems, neurological effects, and birth defects.	[18]
Fluoride	An appropriate amount of fluoride is necessary for maintaining optimal dental health, yet excessive levels can result in tooth discolouration, pitting, and complications related to joints and bones.	[18, 19]
Iron	As per the World Health Organization (WHO), iron is not considered a direct health hazard in water. It is an essential nutrient without specific health-based guidelines. Nevertheless, increased levels of iron can result in taste concerns, as highlighted by Gray (1999).	[15, 18]
Aluminum	There are no adverse effects on health. The Centres for Disease Control and Prevention (CDC) notes that it has only an aesthetic impact, adding color to the water.	[20]
Chromium	Skin irritation is caused by allergies, as reported by both the Environmental Protection Agency (EPA) and the Centres for Disease Control and Prevention (CDC).	[20, 21]
Nitrates	The Environmental Protection Agency (EPA) cautions that infants who ingest elevated levels of these substances for a prolonged period may experience life-threatening consequences if not treated.	[21]

This contamination not only impacts public health but also has economic repercussions, such as lost productivity and increased medical costs. Moreover, it hinders educational outcomes, particularly for children who are most vulnerable to its effects [16]. According to the study by Hunter et al. (2010), investing 1 USD in safe drinking water infrastructure can save the economy between 5 to 45 USD by reducing water-associated illnesses [17].

5.3 REMOVAL OF POLLUTANTS FROM DRINKING WATER BY THE ELECTROCOAGULATION PROCESS

EC is an effective approach for removing pollutants from drinking water. It involves using an electric current to destabilize and coagulate contaminants like organic compounds, suspended solids, and heavy metals. This process is particularly useful for treating water with high turbidity or complex pollutant mixtures. This process improves water quality by reducing pollutants to safe levels, making it a valuable technique for ensuring clean and safe drinking water. Its advantages include simplicity, scalability, and the ability to operate without the need for chemical additives in some cases.

5.3.1 Arsenic Removal

Arsenic can seep into drinking water through natural reservoirs in the ground or due to industrial or agricultural activities. According to the EPA US 2006 standards, with a maximum contamination level of 0.01 mg/L, arsenic can pose a significant health risk globally, notably in countries like India and Bangladesh, where arsenic contamination has led to widespread poisoning cases [22]. Exposure to As in drinking water can result in arsenicosis, a disease characterized by skin thickening, discoloration, gastrointestinal issues like stomach pain, nausea, vomiting, and diarrhea, as well as neurological symptoms such as numbness, partial paralysis, and even blindness. Moreover, arsenic is associated with various types of cancers, including skin, bladder, lung, kidney, nasal passages, liver, and prostate cancers. The toxicity and elimination of arsenic compounds and their metabolites are significantly influenced by their oxidation states, namely Arsenite (As III) and Arsenate (As V), and the degree of methylation of arsenicals. Studies indicate that As(III) exhibits ten times greater toxicity than As(V) [23]. Kobya et al. (2011) found that the EC method effectively reduced the remaining concentration of As to less than 10 μg/L, which is below the WHO acceptable limit. The optimum CD for arsenic removal was found to be 0.25 mA/cm^2, with an optimal pH value of 6.5 and 7.0 for Fe and Al electrodes, respectively. Additionally, the recommended electrolysis period was observed at 12.5 minutes and 15 minutes for Fe and Al electrodes, respectively [7].

5.3.2 Fluoride Removal

Fluoride is a prevalent element, comprising approximately 5% of the Earth's surface and occurring naturally in trace amounts in numerous water bodies. Excessive fluoride in the environment primarily originates from natural resources and industrially generated pollutants. Natural sources include fluoride-rich rocks and soil that contribute fluoride to groundwater and surface water through water interactions. This leads to higher contamination levels in groundwater compared to surface water due to the leaching of fluoride-bearing rocks during water percolation into the Earth. Factors like evaporation rates, aquifer residence time, and intensive irrigation influence the amount of fluoride added to groundwater.

Fluoride, typically found as a greenish diatomic gas (F_2), is rarely present in pure gaseous form except in specific industrial processes. Industries such as chemical, glass, ceramic, coal power, semiconductor manufacturing, electroplating, brick and iron production, aluminum smelting, and beryllium extraction are major contributors to fluoride-rich effluents. Additionally, fluoride enters the human body through everyday products like toothpaste, mouthwash, medications, cosmetics, and chewing gum [24]. Various research studies have documented the occurrence of fluoride in groundwater across multiple countries located within the five fluoride belts. These countries include Pakistan, Sri Lanka, South Korea, Turkey, Iran, India, China, the United States, Mexico, Sweden, Spain, and Argentina, as well as several regions in Africa [25]. According to López-Guzmán M. et al. (2019), a prototype EC cell demonstrates higher efficiency for fluoride removal from well water. The initial fluoride content in the well water is 5 mg/L. Under optimal working conditions,

CD of 4.5 mA/cm², pH of 5, and treatment period of 15 minutes, the fluoride removal rate reaches 85.68%, resulting in a final concentration that is below the recommended level set by the WHO. The acidic pH of the water facilitates easier fluoride removal, achieving a removal rate of 94.98% and reducing the fluoride concentration to less than 1 mg/L. Conversely, Castañeda and Nava (2022) noted that the EC technique exhibits a higher effectiveness in the removal of As and fluoride from groundwater. Initially, the groundwater assessment showed that the concentration of fluoride was at 1.65 mg/L and As at 0.026 mg/L. Both Al and Fe plates were employed as sacrificial anodes in the EC reactor, generating coagulants comprising aluminum and iron species during the process. Their findings reveal an effective elimination of As and F from deep well water using a hybrid electrode made of Al and Fe materials. The final concentration of As and fluoride decreased to $C_{As} = 0.0002$ mg/L and $C_{F-} = 0.88$ mg/L, respectively, meeting the acceptable limits set by the WHO, at an optimum CD of 5 mA/cm² [7].

5.3.3 Iron Removal

The existence of Fe in water is a common issue faced by consumers and water professionals, ranking closely after water hardness. Groundwater globally often contains iron, with reports of severe contamination in India by various states, including Assam, Orissa, Chhattisgarh, Andhra Pradesh, and Karnataka. Some localized areas with high Fe levels are also noted in states such as Bihar, Jharkhand, Madhya Pradesh, Kerala, Maharashtra, Northeastern states, Uttar Pradesh, Rajasthan, Tamil Nadu, and Punjab. Although Fe in drinking water does not pose direct health risks, it is considered undesirable due to aesthetic concerns like high turbidity, discoloration, bad taste, and staining. The WHO recommends Fe levels in drinking water to be below 0.3 mg/L, while the European Commission suggests a limit of less than 0.2 mg/L. In India, the discharge limit for Fe is set at 0.3 mg/L [26].

Iron typically exists in two oxidation states: the reduced soluble divalent ferrous form (Fe^{2+}) and the oxidized trivalent ferric form (Fe^{3+}). Various methods have been utilized for removing iron from groundwater, including oxidation-precipitation-filtration, lime softening, ion exchange, adsorption, bioremediation, and membrane processes. Among these, the most commonly employed methods for Fe removal are oxidation-precipitation-filtration, ion exchange, lime softening, and membrane process. Recent studies have shown that electrochemistry presents a promising substitute for traditional methods for wastewater treatment. One of these electrochemical techniques is electrochemical coagulation, which involves the electrochemical generation of destabilizing agents that facilitate charge neutralization for the removal of pollutants. This method has gained considerable traction in water and wastewater treatment applications. EC offers several benefits, including effective removal of particles, a compact treatment setup, cost-effectiveness, and the potential for full automation. It is known for minimal sludge generation, reduced chemical usage, and ease of operation [27].

5.3.4 Microplastic Removal

Since plastic has gained significant attention from the public, its global production has surged over the past century, increasing from 1.7 million tons in the 1950s to

288 million tons in 2012, and then continued to rise to 361 million tons by 2019. Plastic has become an integral part of modern life, extensively utilized in everyday items such as food packaging films, shopping bags, disposable containers, clothing hangers, storage bins, and even in advanced applications like optical lenses, artificial organs, and aerospace components. The extensive manufacturing and utilization of plastics, along with insufficient waste disposal methods, have led to a significant buildup of plastic waste and remnants, which in turn has played a role in the rise of microplastics (MPs). MPs are categorized into primary and secondary based on their origin and usage. Primary MPs are deliberately created at microscopic scales for particular industrial or domestic uses like microbeads commonly seen in detergents, cosmetics, and personal care products such as facial cleansers, body washes, toothpaste, and scrubs. Conversely, secondary MPs are generated when larger plastic items undergo fragmentation due to mechanical stress, abrasion, photolysis, weathering, and microbial degradation [28].

Numerous studies have documented and foreseen the release of MPs from both primary and secondary sources. Furthermore, several reports highlight that effluents from wastewater treatment plants are significant contributors to MP discharge and a potential source of MPs. According to studies, MPs are commonly found in air, sediment, and notably in aquatic environments (as shown in Figure 5.3). Research on the distribution of MPs across ocean gyres, polar and deep-sea regions, freshwater systems acting as temporary repositories, and in tap and bottled water has consistently detected traces of MPs. In some cases, these traces have been found in concentrations reaching significant levels. In previous studies, it was found that the anode made of Fe and Al materials was effective for the degradation of MPs from water and wastewater. Shen et al. (2022) conducted an EC study involving Fe and Al anodes and investigated the removal of four common MPs (with particle sizes ranging from 6.3 to 286 μm). Their findings reveal that the Al anode exhibits higher MP degradation efficiency than the Fe anode. Furthermore, to avoid passivation effects resulting from polarity inversion, it is recommended to employ similar materials for both electrodes (e.g., the anode and the cathode).

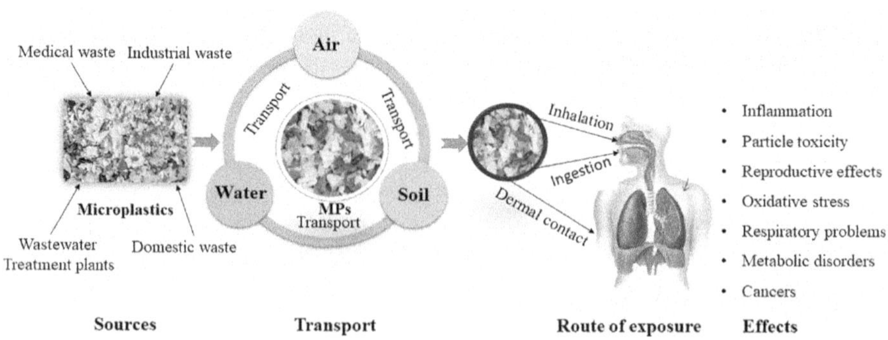

FIGURE 5.3 Pathways of human exposure to microplastic pollution.

5.3.5 Bacteria and Viruses

In drinking water, the removal of bacteria and viruses is crucial to ensure water safety. Various methods such as filtration, disinfection (e.g., chlorination and UV treatment), membrane separation, and EC processes are employed for this purpose. Filtration systems effectively capture bacteria and viruses, while disinfection methods like chlorination destroy their viability. UV treatment is particularly effective against viruses by disrupting their genetic material. Overall, a combination of these techniques is often used to achieve comprehensive removal of bacteria and viruses in drinking water treatment plants [29]. Further research is needed to explore the removal of viruses and pathogenic bacteria, especially concerning the extensive application of the EC process [30].

Anfruns-Estrada et al. (2017) used an EC study with an Fe/Fe electrode configuration to remove bacteria at a CD of 20 mA/cm^2 and natural pH. In 60 minutes, Escherichia coli was reduced by 5 log units, resulting in <1 bacteria/mL. Coliphages and eukaryotes were also removed. Katherine and her colleagues (2014) conducted a study about targeting bacteria using an EC method. They used a soluble Al electrode in E. coli cultures within model solutions and carried out two different experiments—one in a batch reactor and the other in a pilot reactor. The results showed that a 97% reduction in bacteria was achieved within 35 minutes at a CD of 0.22 A. Compared with chemical coagulation, they noted a 2-log factor greater bacteria reduction, using equivalent amounts of Al, in the EC process. In addition to the stand-alone EC process, integrating EC with other purification methods has shown promise in the effective degradation of bacteria and viruses. Conversely, Tanneru et al. (2012) revealed a 1.0 to 1.5-log reduction in viruses from surface water by employing a hybrid EC-MF system at a higher Fe dosage of 13 mg/L [7].

5.4 SLUDGE TREATMENT AND VALORIZATION

EC effectively eliminates impurities from water, yet it produces gases and sludge that may be deemed as waste, emphasizing the significance of research on sludge valorization. The potential applications of EC sludge remain largely unexplored, with only a limited number of studies investigating its utilization or valorization in specific domains. Several studies have explored the utilization of sludge in various applications such as adsorbents, fertilizers, catalysts, construction materials, and pigments, along with some research focusing on EC sludge valorization [31].

5.4.1 Waste Adsorbents

On the contrary, the significant generation of sludge as waste from treatment plants and its subsequent management pose sensitive environmental challenges. Therefore, there is a growing need for innovative research to sustainably treat sludge as a valuable resource. One promising approach is to transform sludge into valuable adsorbents for removing pollutants, such as EC sludge, which can be repurposed into alternative value-added products like adsorbents. This approach is beneficial for optimizing resource usage and managing secondary pollution effectively, as it tackles both the challenge of underutilized resources and the significant disposal issues linked with sludge. Using recyclable EC sludge to prepare adsorbents, the two objectives of maximizing resource utilization and enhancing environmental stewardship can be achieved. For example, iron-based adsorbents developed from EC sludge offer a cost-efficient solution with impressive adsorption capacities, especially in the removal of F– ions because of their unique chemical attributes and thermal stabilities. EC sludge mainly consists of iron oxide-hydroxide precipitates produced in redox reactions during wastewater treatment. While most reports on de-fluoridation indicate the use of commercial iron oxide or composites, their costs may not be sustainable or affordable for communities compared to adsorbents derived from EC sludge. Hence, the direct application of untreated EC sludge or adapted adsorbents in diverse pollutant adsorption applications has attracted considerable interest in the realms of material recycling, resource optimization, and industrial wastewater treatment [32].

Golder et al. (2006) carried out a study on red dye removal from a synthetic solution having chromium sulfate using EC-generated sludge. Sludge was turned into an adsorbent by drying it at 105°C for 12 hours and grinding it into powder. The ideal mixing speed in the adsorption trial was 200 rpm. The efficiency of dye removal was found to be higher at lower pH levels. In addition, the removal rate increased with lower initial dye concentrations and higher dosages of the adsorbent. At a pH of 3, the highest adsorption capacity was observed, reaching a peak value of 513 mg/g. EC sludge generated through processes utilizing Al electrodes exhibited efficient adsorption characteristics for red dye [33].

In another study, Golder et al. (2006) looked at the possibility of using EC sludge as an adsorbent for the removal of phosphate. For this purpose, they utilized a synthetic solution that involves chrome sulfate and EC sludge produced by an EC procedure using Al electrodes. The sludge underwent calcination at 600 °C followed by grinding into a powdered form. The investigation assessed the effects of pH, adsorbent dosage, initial adsorbent concentration, and stirrer speed on the removal of phosphate. Their findings indicated that stirrer speed had minimal to no effect on the removal rate, while higher initial adsorbent concentrations and dosages led to increased removal rates of initial phosphate. Conversely, electrocoagulated metal hydroxides are effective adsorbents for phosphate elimination according to the researchers' findings [34].

5.4.2 Manufacturing Material

There has been a discernible change in the last ten years toward ecologically friendly building materials, the creation of which needed significant work from numerous producers. The European Union has imposed controls on waste materials and their valorization. Several research has looked into the usage of sludge produced from various sources as building materials [35], which act as the fundamental building blocks for many types of constructions (like houses). Nevertheless, the scope of research on the valorization of EC sludge for building applications is quite small. In 2016, Sharma and Josh conducted a study on the utilization of EC-generated sludge as a partial replacement for non-structural building materials. The sludge was produced under specific operating conditions, including a pH of 7.8, an inter-electrode spacing of 22 mm, a CD of 154.32 A/m^2, and an electrolysis period of 135 minutes.

The resulting EC sludge was then dried in an oven at 100°C. EC sludge particles were larger than cement particles when 0–15% sludge was used to reduce cement in building blocks. Compressive strength tests showed decreased strength when EC sludge exceeded 7.5%. Blending EC sludge with cement is useful for building materials, with the ideal proportion being 7.5% [36].

Adyel et al. (2013) studied the use of electrocoagulated metal hydroxide sludge (EMHS) derived from EC treatment of textile industry effluent as a component in manufacturing building materials. The EMHS primarily consisted of iron (87.1%), magnesium (7.3%), silicon (4.8%), and calcium (0.5%), with traces of other elements. The EMHS was obtained from wet materials sourced from the textile industry, while the clay material was sourced from a local brickyard. The EMHS content in the blocks ranged from 10% to 40%, and multiple rounds of testing were conducted for each EMHS composition. Additionally, bricks made of clay, EMHS, and soil were tested. The bricks produced during the tests exhibited satisfactory size and shape, with 30% of the EMHS samples demonstrating appropriate weight characteristics. Bricks that absorbed water were classified as second class, while only 10% of the EMHS samples met the criteria for first-class bricks in compressive strength tests. This study indicates the potential valorization of EMHS as a construction material, particularly for non-load-bearing applications. EC sludge shows promise as a viable ingredient in non-structural construction materials. However, this discovery is just one avenue for valorizing sludge within the construction sector, prompting the need for further exploration into its utilization in various construction materials [37].

5.4.3 Composting Material

Producing composting materials from electrocoagulation-generated sludge is an innovative approach that not only addresses waste management but also contributes to sustainable agricultural practices and environmental conservation. EC is a water and wastewater treatment process that involves the use of electrically generated coagulants to remove contaminants, resulting in the generation of sludge rich in organic and inorganic materials. The chemical composition and characteristics of untreated and digested sewage sludge are illustrated in Table 5.3.

TABLE 5.3
The chemical composition and characteristics of untreated and digested sewage sludge (Reproduced from Chen et al. (2020) [38]).

Chemical composition	Untreated sludge	Digested sludge
pH	5–8	6.5–7.5
Total dry solids (TS), %	2–8	6–12
Nitrogen (N, % of TS)	1.5–4	1.6–6
Phosphorus (P_2O_5, % of TS)	0.8–2.8	1.5–4
Potash (K_2O, % of TS)	0–1	0–3
Protein (% of TS)	20–30	15–20

An essential aspect of sustainable development involves extracting nutrients from wastewater and sludge. Urban and industrial wastewater serve as valuable sources for recovering biologically active components. Sewage sludge obtained from wastewater treatment plants contains a wealth of biomolecules, including nutrients like NPK, antioxidants, lipids, proteins, enzymes, carbohydrates, vitamins, nitrogen, phosphates, minerals, and ammonium compounds [38]. Converting this sludge into composting materials presents several benefits and challenges. Firstly, EC-generated sludge contains a mix of organic matter, nutrients, and trace elements that are beneficial for soil health and plant growth. The organic fraction of the sludge, including suspended solids, biodegradable compounds, and microbial biomass, serves as a valuable source of carbon and nitrogen for composting. These nutrients, along with essential minerals like phosphorus, potassium, and micronutrients, contribute to the fertility of composted materials. Moreover, composting EC-generated sludge helps in recycling nutrients and reducing waste disposal. Instead of disposing of the sludge in landfills or incinerating it, which can have environmental impacts, converting it into compost adds value by creating a nutrient-rich soil amendment. This aligns with the principles of circular economy and sustainable waste management, promoting resource efficiency and reducing the environmental footprint of wastewater treatment processes. However, there are challenges associated with composting EC-generated sludge that need to be addressed. One challenge is the potential presence of heavy metals and other contaminants in the sludge, which may require pretreatment or monitoring to ensure the safety and quality of the composting materials. Effective monitoring and control measures are essential to comply with regulatory standards and prevent any adverse effects on soil, water, or plant health. Another challenge is the proper management of composting processes, including optimizing the carbon-to-nitrogen ratio, maintaining adequate moisture levels, and ensuring proper aeration to facilitate microbial decomposition. The composition of EC-generated sludge can vary depending on the source and treatment process, requiring tailored composting strategies and monitoring techniques for optimal results [39].

5.5 COMPARISON OF ELECTROCOAGULATION WATER TREATMENT PLANT WITH ELECTRODIALYSIS

In 1906, A.E. Dietrich received a patent for his invention of the EC technique, which was first used to treat ship bilge water. EC is a straightforward and effective technique that does not require the addition of coagulants, thereby reducing sludge production. The treatment stages for the EC process have already been discussed in the Introduction section. The key factors influencing the EC process include the surface area of electrodes, the electrode materials, current density, solution pH, and the mixing process during coagulation.

In contrast, according to the IUPAC, electrodialysis (ED) is an electrochemical separation method that uses semi-permeable or ion-selective membranes to remove ions while being subjected to an electric field. The DC power supply which facilitates ion movement; electrodes (where oxidation/reduction occurs); membranes which facilitate counterion transportation and prevent co-ion transport; solvent (which fills the space between the electrode and the membrane to act as a continuous medium for ion transportation); and electrolyte (the current carrier between the cathode and the anode) are the five components needed for a basic ED arrangement [40]. ED is a commonly utilized separation method that has been applied to the elimination of both anions and cations. Anions move toward the anode, while cations migrate to the cathode under the influence of the electric field. Among the industrial uses, ED is utilized for the extraction of metals from water and wastewater and the creation of drinkable water from brackish water. The working principle of ED relies on the voltage gradient across a sequential arrangement of cation and anion exchange membranes positioned between an anode and a cathode. The incoming solution, comprising both positively and negatively charged ions, is introduced into the membrane stack where an electrical voltage is enforced. This leads to the migration of ions toward their respective electrodes. Cation exchange membranes facilitate cation transfer while hindering anion transfer, whereas anion exchange membranes facilitate anion transfer while hindering cation transfer. An ionic solution circulates around the electrodes to sustain the conductivity of the membrane stack while preventing other potentially corrosive ions in the feed solution from reaching the electrodes [41].

5.6 LIMITATIONS AND FUTURE RECOMMENDATIONS

The EC process has various limitations such as regular replacement of sacrificial anodes which is essential to maintain the optimal performance of the process. These anodes degrade over time due to electrochemical reactions, losing their effectiveness in removing contaminants. Cathode passivation is a phenomenon where a layer of insulating material forms on the cathode surface during the EC process. This layer inhibits the flow of electric current and reduces the efficiency of the process. To mitigate passivation effects, regular cleaning or maintenance of the cathode is necessary to ensure sustained operational efficiency in EC systems. Moreover, EC is a highly effective technique for removing various contaminants from water, but it may not eliminate all types of pollutants. Some complex organic compounds or very fine particles may require additional treatment methods in conjunction with EC for complete removal. Another limitation of this treatment process is high energy consumption due to the larger electrode surface area, high current density, and the selected electrode material. Larger electrode surface areas require more electrical power to maintain the desired current density for effective coagulation and contaminant removal. Moreover, higher current densities are often needed to achieve faster coagulation and settling of contaminants, leading to increased power consumption. Similarly, the choice of electrode material can impact power consumption. For instance, using higher-quality electrodes like titanium can reduce power requirements compared to using materials like iron.

Researching and developing specialized electrode materials tailored to target specific contaminants (e.g., microplastics, pharmaceuticals, and perfluoroalkyl substances), improving their adsorption and removal during EC processes and exploring advanced oxidation processes (AOPs) in conjunction with EC are methods to enhance the degradation and removal of persistent contaminants like pharmaceuticals and perfluoroalkyl substances. The implementation of a hybrid system such as EC-MF may achieve higher removal efficiencies for microplastics and smaller molecules.

Furthermore, the development of energy-efficient EC systems may reduce power consumption while maintaining high treatment efficacy. These strategies include optimizing electrode designs for enhanced current distribution, utilizing renewable energy sources like solar or wind power, implementing smart control systems to adjust operating parameters based on real-time conditions, and integrating energy recovery mechanisms such as the electrochemical regeneration of coagulants.

5.7 CONCLUSION

Accessing safe drinking water these days is hard and scarce; water pollution can harm aquatic ecosystems and freshwater supplies. However, more potent and efficient drinking water treatment methods are required to address the risks posed by pollutants in the environment, such as high concentrations of pollutants. EC technique may eliminate a wide range of water pollutants, including organic matter, suspended solids, and metals and metalloids. Moreover, to create a safe environment, EC-generated sludge disposal management to convert the sludge into value-added products like waste adsorbents, manufacturing materials, and composting material is discussed. Furthermore, a comparison study of EC and ED is performed by analyzing the advantages of EC over the ED process for treating drinking water. However, we anticipate that electrochemical methods will become the preferred option for pollutant removal globally soon, once they are fully developed over the next decade.

REFERENCES

[1] United Nations, *The Sustainable Development Goals Report 2022*, United Nations, New York, NY, USA, 2022.

[2] WHO, *World Health Statistics 2022 (Monitoring Health of the SDGs)*, WHO, Geneva, Switzerland, 2022. ISBN 9789240051140.

[3] Y. Mao, Y. Zhao, S. Cotterill, Examining current and future applications of electrocoagulation in wastewater treatment, *Water* 15 (2023). https://doi.org/10.3390/w15081455.

[4] M. Ingelsson, N. Yasri, E.P.L. Roberts, Electrode passivation, faradaic efficiency, and performance enhancement strategies in electrocoagulation—a review, *Water Res.* 187 (2020) 116433. https://doi.org/10.1016/j.watres.2020.116433.

[5] A. Almukdad, M.A. Hafiz, A.T. Yasir, R. Alfahel, A.H. Hawari, Unlocking the application potential of electrocoagulation process through hybrid processes, *J. Water Process Eng.* 40 (2021) 101956. https://doi.org/10.1016/j.jwpe.2021.101956.

[6] D.T. Moussa, M.H. El-Naas, M. Nasser, M.J. Al-Marri, A comprehensive review of electrocoagulation for water treatment: Potentials and challenges, *J. Environ. Manage.* 186 (2017) 24–41. https://doi.org/10.1016/j.jenvman.2016.10.032.

[7] M. Bharti, P.P. Das, M.K. Purkait, A review on the treatment of water and wastewater by electrocoagulation process: Advances and emerging applications, *J. Environ. Chem. Eng.* 11 (2023) 111558. https://doi.org/10.1016/j.jece.2023.111558.

[8] P.P. Das, M. Sharma, M.K. Purkait, Recent progress on electrocoagulation process for wastewater treatment: A review, *Sep. Purif. Technol.* 292 (2022) 121058. https://doi.org/10.1016/j.seppur.2022.121058.

[9] J. Fawell, M.J. Nieuwenhuijsen, Contaminants in drinking water, *Br. Med. Bull.* 68 (2003) 199–208. https://doi.org/10.1093/bmb/ldg027.

[10] C. Binnie, M. Kimber, A. Water, *Basic Water Treatment* (4th ed.), 1801, p. 648.

[11] S. Chowdhury, M.A.J. Mazumder, O. Al-attas, T. Husain, Science of the total environment heavy metals in drinking water : Occurrences, implications, and future needs in developing countries, *Sci. Total Environ.* 569–570 (2016) 476–488. https://doi.org/10.1016/j.scitotenv.2016.06.166.
[12] WHO, Guidelines for drinking-water quality fourth edition incorporating the first addendum, *WHO Chronicle* 38(4) (2017) 104–108.
[13] A. Manual, CAWST Training, *Introduction to Drinking Water Quality Testing,* Centre for Affordable Water and Sanitation Technology, 2009.
[14] A.N. Jones, J. Bridgeman, A fluorescence-based assessment of the fate of organic matter in water treated using crude/purified Hibiscus seeds as coagulant in drinking water treatment, *Sci. Total Environ.* 646 (2019) 1–10. https://doi.org/10.1016/j.scitotenv.2018.07.266.
[15] N. Gray, *Water Technology* (3rd ed.), CRC Press, 2010. https://doi.org/10.1201/9781315276106
[16] D. Bradley, J. Vegh, C.S. Lee, R. Cao, L. Chan, Assessment of water purification technologies in developing countries, *Portl. Int. Conf. Manag. Eng. Technol.* (2018) 1–6. https://doi.org/10.23919/PICMET.2018.8481981.
[17] P.R. Hunter, A.M. MacDonald, R.C. Carter, Water supply and health, *PLoS Med.* 7 (2010). https://doi.org/10.1371/journal.pmed.1000361.
[18] WHO Chronicle, Fourth edition, Guidelines for drinking-water quality, *WHO Chronicle* 38(4) (2011) 104–108.
[19] Shankar B. Kausley, et al., Clean water for developing countries: Feasibility of different treatment solutions, in: *Reference Module in Earth Systems and Environmental Sciences*, Elsevier, 2019: pp. 643–652.
[20] B. Huber, O. Schmid, L. Kilcher, Standards and regulations, in: *In the World of Organic Agriculture*, Routledge, 2010: pp. 59–70.
[21] US Environmental Protection Agency, *2018 Edition of the Drinking Water Standards and Health Advisories Tables*, 2018.
[22] M.M.H. Khan, F. Sakauchi, T. Sonoda, M. Washio, M. Mori, Magnitude of arsenic toxicity in tube-well drinking water in bangladesh and its adverse effects on human health including cancer: Evidence from a review of the literature, *Asian Pacific J. Cancer Prev.* 4 (2003) 7–14.
[23] S. Sharma, A. Bhattacharya, Drinking water contamination and treatment techniques, *Appl. Water Sci.* 7 (2017) 1043–1067. https://doi.org/10.1007/s13201-016-0455-7.
[24] S. Ahmad, R. Singh, T. Arfin, K. Neeti, Fluoride contamination, consequences and removal techniques in water: A review, *Environ. Sci. Adv.* 1 (2022) 620–661. https://doi.org/10.1039/d1va00039j.
[25] L.F. Castañeda, J.F. Rodríguez, J.L. Nava, Electrocoagulation as an affordable technology for decontamination of drinking water containing fluoride: A critical review, *Chem. Eng. J.* 413 (2021). https://doi.org/10.1016/j.cej.2020.127529.
[26] D. Ghosh, H. Solanki, M.K. Purkait, Removal of Fe(II) from tap water by electrocoagulation technique, *J. Hazard. Mater.* 155 (2008) 135–143. https://doi.org/10.1016/j.jhazmat.2007.11.042.
[27] K.S. Hashim, R. AlKhaddar, A. Shaw, P. Kot, D. Al-Jumeily, R. Alwash, M.H. Aljefery, Electrocoagulation as an eco-friendly river water treatment method, in: R. AlKhaddar, R.K. Singh, S. Dutta, M. Kumari (Eds.), *Advances in Water Resources Engineering and Management*, Springer Singapore, Singapore, 2020: pp. 219–235.
[28] S.H. Na, M.J. Kim, J.T. Kim, S. Jeong, S. Lee, J. Chung, E.J. Kim, Microplastic removal in conventional drinking water treatment processes: Performance, mechanism, and potential risk, *Water Res.* 202 (2021) 117417. https://doi.org/10.1016/j.watres.2021.117417.

[29] H. Jain, R. Dhupper, A. Shrivastava, M. Kumari, Enhancing groundwater remediation efficiency through advanced membrane and nano-enabled processes: A comparative study, *Groundw. Sustain. Dev.* 23 (2023) 100975. https://doi.org/10.1016/j.gsd.2023.100975.
[30] D. Ghernaout, Virus removal by electrocoagulation and electrooxidation: New findings and future trends, *J. Environ. Sci. Allied Res.* 02 (2019) 37–43. https://doi.org/10.29199/2637-7063/esar-202024.
[31] K. Rajaniemi, S. Tuomikoski, U. Lassi, Electrocoagulation sludge valorization—a review, *Resources* 10 (2021). https://doi.org/10.3390/resources10120127.
[32] T.A. Aragaw, Recycling electro-coagulated sludge from textile wastewater treatment plants as an adsorbent for the adsorptions of fluoride in an aqueous solution, *Heliyon* 7 (2021) e07281. https://doi.org/10.1016/j.heliyon.2021.e07281.
[33] A.K. Golder, A.N. Samanta, S. Ray, Anionic reactive dye removal from aqueous solution using a new adsorbent-sludge generated in removal of heavy metal by electrocoagulation, *Chem. Eng. J.* 122 (2006) 107–115. https://doi.org/10.1016/j.cej.2006.06.003.
[34] A.K. Golder, A.N. Samanta, S. Ray, Removal of phosphate from aqueous solutions using calcined metal hydroxides sludge waste generated from electrocoagulation, *Sep. Purif. Technol.* 52 (2006) 102–109. https://doi.org/10.1016/j.seppur.2006.03.027.
[35] J. Balasubramanian, P.C. Sabumon, J.U. Lazar, R. Ilangovan, Reuse of textile effluent treatment plant sludge in building materials, 26 (2006) 22–28. https://doi.org/10.1016/j.wasman.2005.01.011.
[36] P. Sharma, H. Joshi, Utilization of electrocoagulation-treated spent wash sludge in making building blocks, *Int. J. Environ. Sci. Technol.* 13 (2016) 349–358. https://doi.org/10.1007/s13762-015-0845-7.
[37] T.M. Adyel, S.H. Rahman, M.M. Zaman, H.M. Sayem, M. Khan, M. Abdul Gafur, S.M.N. Islam, Reuse feasibility of electrocoagulated metal hydroxide sludge of textile industry in the manufacturing of building blocks, *J. Waste Manag.* 2013 (2013) 1–9. https://doi.org/10.1155/2013/686981.
[38] J. Chen, J. Li, X. Zhang, Z. Wu, Pretreatments for enhancing sewage sludge reduction and reuse in lipid production, *Biotechnol. Biofuels* 13 (2020) 1–10. https://doi.org/10.1186/s13068-020-01844-3.
[39] T.C. Sichler, C. Adam, D. Montag, M. Barjenbruch, Future nutrient recovery from sewage sludge regarding three different scenarios—German case study, *J. Clean. Prod.* 333 (2022) 130130. https://doi.org/10.1016/j.jclepro.2021.130130.
[40] H.K. Hansen, C. Gutiérrez, J. Leiva Gonzalez, A. Lazo, M.E. Hansen, P. Lazo, L.M. Ottosen, R. Ortiz, Combined electrodialysis and electrocoagulation as treatment for industrial wastewater containing arsenic and copper, *Membranes* 13 (2023). https://doi.org/10.3390/membranes13030264.
[41] I. Ali, T.A. Khan, M. Asim, Removal of arsenic from water by electrocoagulation and electrodialysis techniques, *Sep. Purif. Rev.* 40 (2011) 25–42. https://doi.org/10.1080/15422119.2011.542738.

6 Application of the Electrocoagulation Process in Industrial Wastewater Treatment

6.1 INTRODUCTION

Industrial wastewater is a significant environmental concern due to its complex composition and potential to cause pollution. Several industries such as textile, steel, pharmaceutical, paper and pulp, tannery, and chemical processing industries generate wastewater containing a wide range of contaminants, including heavy metals, organic compounds, and suspended solids (SS). Traditional wastewater treatment methods often struggle to effectively remove these contaminants, leading to the need for innovative and efficient treatment technologies. One such technology that has gained attention for its applicability in industrial wastewater treatment is electrocoagulation [1]. EC is an electrochemical method that involves metal electrodes (generally made of iron and aluminum) to destabilize and remove contaminants from water. It involves the generation of coagulant species through the electrolysis of metal electrodes, causing the generation of metal hydroxide flocs that aid in the precipitation and removal of pollutants.

The application of EC in industrial effluent purification offers several advantages over conventional methods. Firstly, EC is effective in removing various types of contaminants including heavy metals, suspended solids, organic compounds, and even pathogens like bacteria and viruses. This versatility makes it suitable for treating wastewater from diverse industrial processes. Secondly, EC does not require the addition of chemicals for coagulation, reducing the need for hazardous chemicals and the associated handling and disposal costs. This aspect aligns with the principles of green chemistry and sustainable wastewater treatment practices, making EC a desirable option for industries aiming to minimize their environmental impact. Thirdly, EC can achieve rapid treatment and high pollutant removal efficiencies, making it suitable for treating large volumes of wastewater generated by industrial operations. The process is scalable and can be customized to meet specific treatment objectives and regulatory requirements. Moreover, EC offers operational flexibility, allowing for adjustments in parameters such as current density, electrode material, pH, and electrolysis period to optimize treatment performance based on the characteristics of the wastewater and the target pollutants. The effectiveness of EC in industrial

DOI: 10.1201/9781003499671-6

effluent purification has been demonstrated in various studies and applications across different industries [2].

For example, in the textile industry, EC has been used to remove dyes and colorants from wastewater, addressing one of the major environmental challenges faced by textile manufacturers [3]. Likewise, in the steel industry, EC technology is utilized to degrade these heavy metals, including chromium (Cr), nickel (Ni), and zinc (Zn) as well as oil and grease from machining operations [4]. Furthermore, in the pharmaceutical and chemical industries, EC has been employed to treat wastewater containing organic pollutants, pharmaceutical residues, and complex chemical compounds. The electrochemical reactions during EC can break down organic molecules and facilitate their removal through flocculation and sedimentation [5]. Similarly, in the tannery industry, wastewater contains organic compounds, fats, oils, greases, and chemicals from leather processing. EC technology is employed to treat tannery wastewater by causing the coagulation of organic matter and the separation of suspended solids. The electrochemical reactions in EC facilitate the removal of colorants and the reduction of organic load, resulting in cleaner effluents that can be safely discharged or reused [6].

6.2 TOXIC CONTAMINANTS IN INDUSTRIAL WASTEWATER AND ITS ENVIRONMENTAL REGULATIONS

Wastewater originating from diverse industrial sectors contains numerous pollutants that pose toxicity and hazardous risks to both human and aquatic life as well as agricultural systems. These contaminants encompass heavy metals such as Cr, Zn, Pb, Cu, Fe, Cd, Ni, As, and Hg. The main contributors to these contaminants are the industries involved in paint and dye manufacturing, textiles, pharmaceuticals, paper production, and fine chemicals [7]. Additionally, industrial wastewater also contains phenol and phenolic compounds, which are significant pollutants primarily discharged by oil refineries, phenol-formaldehyde resin manufacturing, and bulk drug industries. Textile industries produce a wide range of compounds, from heavy metals like Cr to surfactants, bleaching agents such as hydrogen peroxide and chlorine, AOX, sodium silicate, and alkaline bases. Perfluoroalkyl acids (PFAAs) are employed as surface protectants due to their high surface activity, stability, and oil-water-repellent properties.

Nevertheless, the two PFAAs perfluorooctane sulfonate (PFOS) and perfluorooctanoic acid (PFOA) pose potential health risks. PFOS primarily originates from textile treatment, metal plating, and semiconductor industries, whereas PFOA is primarily discharged by the fluoropolymer production and processing sectors. These substances are predominantly dispersed through the wastewater discharged from these industrial operations. In addition to these contaminants, the elevated salinity levels in wastewater also pose numerous negative impacts on various aspects of life. Consequently, the removal of salt from wastewater has gained equal significance to the elimination of organic matter (OM) and other pollutants in numerous countries. Industries such as petroleum, leather, food processing, and agro-based sectors contribute to the generation of high-salinity wastewater, primarily composed of NaCl [8]. The pollutants discharged from some major industries are summarized in Table 6.1.

TABLE 6.1
Major water pollutants from various industrial effluents.

Industry	Major contaminants	Environmental regulations		References
		Parameters	**Standards limiting concentration in mg/L, except for pH and color**	
Textile	Iron, chromium, chlorinated compounds, urea, salts, hydrogen peroxide, high-pH NaOH, surfactants	pH	6–10	[9]
		Biochemical oxygen demand (BOD)	80–6000	
		Chemical oxygen demand (COD)	150–12000	
		Oil and grease	10–30	
		Total suspended solids (TSS)	15–8000	
		Free ammonia	<10	
		Total dissolved solids (TDS)	2900–3100	
		Chloride	1000–1600	
		Sodium	70	
		Silica	<15	
		Color (Pt-Co)	50–2500	
Pharmaceutical	Cadmium, nickel, phenolic compounds	pH	6–8.5	[10]
		Oil and grease	10	
		BOD	100	
		COD	250	
		TSS	100	
		Mercury	0.01	
		Arsenic	0.2	
		Chromium	0.1	
		Lead	0.1	
		Cyanide	0.1	
		Phenolics	1.0	
		Sulfides	2.0	
		phosphate	5.0	
Steel (Integrated Steel and Iron)	Heavy metals (lead, arsenic, chromium), oil and grease, cyanide, solid waste, and slags	pH	6–8.5	[11]
		TSS	100	
		BOD	30	
		COD	250	
		Oil and grease	10	
		Ammoniacal nitrogen (as N)	50	
		Cyanide (as CN^{+6})	0.2	
		Phenol	1.0	

(Continued)

TABLE 6.1 *(Continued)*
Major water pollutants from various industrial effluents.

Industry	Major contaminants	Environmental regulations		References
		Parameters	Standards limiting concentration in mg/L, except for pH and color	
Tannery	Chromium, sulfides, Formaldehyde, salts, organic solvents	pH	6–9	[12]
		BOD	20	
		COD	250	
		TSS	50	
		TDS	2100	
		Sulfides (as S)	2.0	
		Total chromium (as Cr)	2.0	
		Hexavalent chromium (as Cr^{+6})	0.1	
		Oil and grease	10	
Paper and pulp	Small pulp and paper industry (capacity below 24000 MT per annum)	pH	5.5–9.0	[13]
		TSS	100	
		BOD	30	
		Absorbable organic halogens (AOX) in effluent discharge	2.0 kg/ton of paper produced	
	Large pulp and paper industry (capacity above 24000 MT per annum) Organic and chlorophe-nolic compounds, suspended solids, AOX, lignin, tannins, sterols, colors, biocides, etc.	pH	7.0–8.5	
		BOD	30	
		COD	350	
		TSS	500	
		AOX	1.0 kg/ton of paper produced	

6.3 REMOVAL OF POLLUTANTS FROM VARIOUS INDUSTRIES BY THE ELECTROCOAGULATION PROCESS

Industries such as steel, textile, tannery, paper and pulp and pharmaceuticals generate wastewater containing pollutants like dyes, heavy metals, organic compounds, and suspended solids. EC is particularly effective in treating these wastewater streams by targeting specific contaminants through electrochemical reactions. Figure 6.1 depicts the application of the EC process in the treatment of various industrial wastewater.

6.3.1 Steel Industry Wastewater

Iron and steel are extensively utilized in construction, machinery, and equipment manufacturing, playing a crucial role in human civilization's advancement. The steel

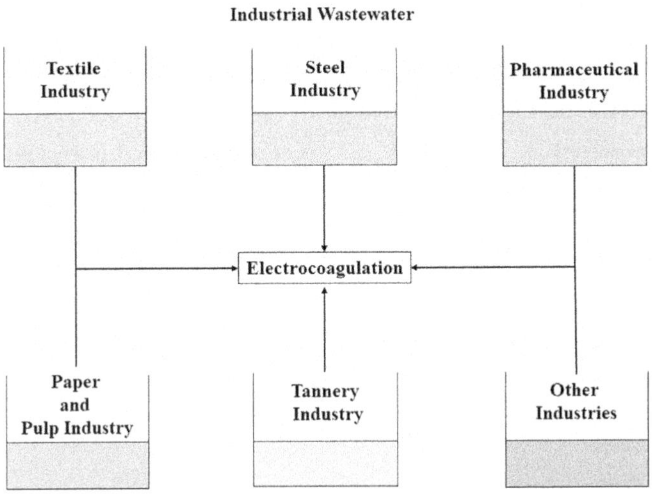

FIGURE 6.1 Application of electrocoagulation in industrial wastewater treatment.

industry generates a significant amount of heavy metal pollutants. These pollutants can enter the human body through inhalation and respiration, leading to various harmful effects. For instance, elevated blood pressure and symptoms of anemia can result from heavy metal exposure. Several treatment methods are used for wastewater in the steel industry, including reverse osmosis, sedimentation, filtration, and electrodialysis. These methods typically involve three main steps: primary, secondary, and tertiary treatment. In addition, EC is another effective treatment method for industrial wastewater, offering the advantage of not requiring any chemicals during the process. Vignesh et al. (2017) treated simulated electroplating wastewater using the EC process along with the scraps of metal as electrodes. Their findings show that the Ni ion degradation efficiency was 95.9% and 94.1% for Fe and Al electrodes, respectively, under operating conditions such as a CD of 30 mA/cm^2 and a treatment period of 60 minutes. Furthermore, Al-Shannag et al. (2015) examined the degradation of heavy metals, including Cu^{2+}, Ni^{2+}, Zn^{2+}, and Cr^{3+}, found in metal plating industry effluent. Their study found more than 97% degradation of heavy metals under operating conditions such as a pH of 9.56, a CD of 4 mA/cm^2, and a treatment time of 45 minutes [11].

6.3.2 Textile Industry Wastewater

The wastewater produced from the textile industry is one of the most heavily contaminated wastewater compared to other industrial effluents. As per the World Bank, around 17% to 20% of industrial effluent originates from textile sectors globally. India, following China as the world's top exporter of textiles, faces significant challenges in managing textile wastewater pollution. Studies conducted in cities like Pali, and Rajasthan, which have numerous textiles dyeing and printing industries, have revealed that the chemical properties often exceed standards set by the Bureau of Indian Standards. It is characterized by COD, salt content, high turbidity, and total organic carbon (TOC) levels. These characteristics are shown in Table 6.2.

TABLE 6.2
Properties of textile industry effluent (Reproduced with permission from Sandhya et al. (2008) @ Elsevier [14]).

Parameters	Concentration, mg/L (except pH)
pH	9.5–12.5
TSS	60–416
TDS	4500–12800
TOC	26390–73190
BOD	25–433
COD	1835–3828
Aromatic amines	20–75
Ammonia	2–3
Chlorides	1200–1375
Sulfate	700–2400

Aquatic ecosystems may suffer significantly when textile effluents are discharged into the environment without prior treatment. These effects include bioaccumulation and the obstruction of light penetration, which impacts aquatic organisms' growth negatively. Stipulated dyes and their breakdown products from textiles can harm aquatic life and human health, causing issues such as intestinal cancer, fetal cerebral abnormalities, allergies, and irritation of mucous membranes, the upper respiratory tract, eyes, and skin. Textile wastewater contains potent dyes that are toxic, highly stable, and persistent pollutants, resistant to conventional water treatment methods. In comparison to the chemical coagulation process, EC stands out as the fastest, most cost-effective, and efficient process for treating textile wastewater [14]. Kobya et al. (2003) conducted a study on treatment of textile effluents using an EC technique involving Al and Fe electrodes. The Fe electrodes demonstrated higher COD removal and lower energy consumption than the Al electrodes. The Fe electrodes exhibit higher activity in a neutral to alkaline environment, with a pH range of 6 to 9. Within this pH range, turbidity and COD elimination efficiencies range from 75% to 98% and 47% to 77%, respectively. However, these efficiencies drop to 87% and 26%, respectively, at pH levels above 10. On the other hand, the Al electrodes are more effective under acidic conditions with a pH below 6. At pH levels below 6, turbidity elimination reaches up to 98% and COD elimination ranges from 61% to 65%, while efficiencies decrease above pH 6. For comparable turbidity or COD elimination efficiencies, the Fe electrodes require a current density of 80 to 100 A/m^2, whereas the Al electrodes need 150 A/m^2 with an electrolysis period of 10 minutes [15]. In addition, Kobya et al. (2014) conducted another study on TOC and turbidity removal using the EC process. They achieved elimination efficiencies of 77% and 94% for TOC and turbidity, respectively, using Fe electrodes in a treatment time of 34 minutes, while the Al electrodes showed 68% and 99% removal in a treatment period of 33.9 minutes, respectively [16]. Furthermore, Tezxan Un and Aytac (2013) utilized an EC reactor, specifically a packed bed type, designed for treating textile wastewater. The reactor, constructed from Fe, operated in a semi-continuous mode. The study found that under operating conditions that include a pH of 9, a CD of 20 mA/cm^2, a flow rate of 20 mL/min, and a treatment period of 10 minutes, the reactor achieved complete turbidity removal and 96.88% COD removal [17]. In a separate study by Kobya and colleagues (2016), a continuous EC approach was used for treating textile effluents. They employed rectangular-shaped electrodes, comprising two anodes and two cathodes made of Fe during the study. They utilized both Fe and Al electrodes simultaneously in a monopolar parallel arrangement. The ideal operating parameters for both electrodes include a CD of 65 A/m^2, an inlet flow rate of 0.010 L/min, and a treatment period of 80 minutes. Under these conditions, COD, TOC, and turbidity degradation efficiencies were 85%, 76%, and 95% for the Fe electrode (as the anode), and 77%, 72%, and 95% for the Al electrode (as the anode). Their findings also highlighted the influence of initial concentration on the degradation efficiency of the process. Concentrations of COD exceeding 2000 mg/L revealed a decrease in efficiency, possibly due to electrode deactivation as a result of deposited dye molecules on the surface of the electrode [18].

6.3.3 Tannery Wastewater

The wastewater produced by the leather industry is significantly polluted with a variety of organic and inorganic contaminants. Details regarding the properties of tannery industry effluent can be found in Table 6.3. Due to its considerable noxiousness, it is imperative to treat this effluent before releasing it into the atmosphere. The EC process is effective at eliminating COD, Cr, TDS, and other contaminants from tannery effluent. Benhadji et al. (2011) conducted a study on tannery effluent treatment using EC along with Al and steel electrodes (Al as the anode and Al/steel as the cathode). The study found that using Al as the cathode resulted in the highest degradation efficiency. They achieved more than 90% removal of BOD, COD, turbidity, Cr, Fe, nitrate, and nitrite. Nitrate ions present in water can be reduced and broken down by

TABLE 6.3
Properties of tannery industry effluent.

	Characteristic value, mg/L (except pH)			
Parameters	**[19]**	**[20]**	**[21]**	**[22]**
pH	7.1	12	8.5	7.6–7.8
COD	7680	25300	322	12225–17618
BOD_5	1326	10850	160	–
Cr	18.4	–	5.7	44–50
Fe	3.12	–	–	3.9–4.4
Zn	0.5	–	–	0.4–0.5
Chloride	284	–	–	–

aluminum oxidation at the anode, which is responsible for the almost total elimination of nitrate and nitrite. This reduction process converts nitrate ions to nitrite ions and subsequently to ammonia and nitrogen gas [19].

$$3NO_3^- + 2Al + 3H_2O \rightarrow 3NO_2^- + 2Al(OH)_3 \quad (6.1)$$

$$NO_2^- + 2Al + 5H_2O \rightarrow NH_3 + 2Al(OH)_3 + (OH)^- \quad (6.2)$$

$$2NO_2^- + 2Al + 4H_2O \rightarrow N_2 + 2Al(OH)_3 + 2OH^- \quad (6.3)$$

Kongjao et al. (2008) employed the EC technique for treating effluent produced by the tannery industry. In their findings, the degradation observed was 100% for Cr, 99% for grease and oil, 62% for TKN, 50% for TDS, and >95% for COD, BOD, and TSS [20].

In a distinct study by Elabbas et al. (2016), chrome tanning wastewater was treated using a batch EC technique by incorporating two different types of Al electrodes, one is pure Al and the other one is Al alloy (Al/Cu/Mg). The study noted that highly concentrated effluents required longer electrolysis times (6 hours) and higher current densities (up to 400 A/m^2) for effective treatment. It was discovered that the dilution of concentrated wastewater increased the effectiveness of treatment and decreased energy usage. Additionally, it was discovered that Duralumin (aluminum alloy) electrodes were more effective than pure Al electrodes due to their higher solubility and greater charge retention capabilities [23]. Conversely, Sengil et al. (2009) used mild steel (MS) and aluminum (Al) electrodes as the anode in a batch EC process to treat effluent from the liming drum tannery industry. Their findings indicated that there was no noticeable difference in the degradation of oil and grease when using the Al and MS electrodes, but the MS electrodes showed a higher removal efficiency for COD and sulfide. Specifically, they achieved 96% removal of grease and oil, 82% removal of COD, and 90% removal of sulfide under optimum conditions of a pH of 3, a current density (CD) of 3.5 mA/cm^2, and an electrolysis period of 10 minutes [20].

6.3.4 Paper and Pulp Industry Wastewater

The paper and pulp industry is characterized as a water-intensive sector and ranks third globally for freshwater consumption. The water utilization in this industry varies depending on the type of paper being manufactured and can reach up to 60 cm^3/ton paper manufactured. Consequently, the industry generates a substantial amount of wastewater containing numerous pollutants. This effluent has been found to include about 500 different types of adsorbable organic halides, including resin acids, dioxins, phenols, catechols, chlorate, furans, and others. Due to the presence of higher concentrations of these substances in paper and pulp industry effluents, it is imperative to treat them to mitigate their potential impacts on the aquatic environment. During the last few decades, the use of the EC process has gained popularity for treating paper industry effluent [24]. Sridhar et al. (2011)

examined a study on the treatment of bleaching effluent from the paper and pulp industry using the EC process. They examined effective degradation for color and COD of more than 90%, along with approximately 87% removal of BOD. Black liquor, a major by-product of the paper and pulp industry, is considered a significant pollutant due to its high lignin content of around 50%, which is non-biodegradable due to its complex chemical structure [25]. In addition, Zaied et al. (2009) treated the black liquor wastewater produced from the paper and pulp industry and found degradation of 99%, 98%, and 92% for color, COD, and polyphenols, respectively [26]. In contrast, Ugurlu et al. (2008) employed EC for treating lignin and phenol, achieving removal rates of 80% for lignin and 98% for phenol using Al electrodes, and 92% and 93%, respectively, using Fe electrodes. The study also noted that increasing CD led to a higher removal efficiency. Furthermore, EC can serve as a pretreatment method for biological processes in treating black liquor wastewater. This technique can remove organic non-biodegradable dissolving compounds by up to 46%, thereby increasing the biodegradability index from 0.11 to 0.46 [27]. Moreover, Mahesh et al. (2006) conducted a comparison study using a continuous EC process for treating paper and pulp industry effluent at varying flow rates. They found COD and color degradation efficiencies to be 78.20% and 79% and 82.15% and 90% for the flow rates of 1.0 and 0.5 dm^3/h, respectively. The total solids (TS) concentration in the wastewater slurry also decreased by 65% after an electrolysis period of 3 hours and a flow rate of 1 dm^3/h [28].

6.3.5 Pharmaceutical Wastewater

The pharmaceutical industry is regarded as a low-volume, high-value asset due to its production of a wide range of compounds using various synthetic materials. Consequently, the wastewater produced by these industries is typically highly toxic. According to the Singh et al. (2016) investigation, treating pharmaceutical wastewater using the EC process results in maximum degradation efficiency of COD and color with the minimum usage of energy. They used Al and stainless steel (SS) electrodes in their study. The Al electrode achieved the highest color removal of 99.88%, while the SS electrode showed the highest COD removal of up to 78.88% [29]. In addition, Veli et al. (2016) treated hospital wastewater using EC with three distinct electrode materials (Fe, Al, and SS) in a batch mode. Among these electrodes, SS demonstrated the highest TOC degradation efficiency of 99.89% at a pH of 7.8 and a CD of 2.68 mA/cm^2 [30]. In contrast, Ahmadzadeh et al. (2018) investigated the degradation of metronidazole (MNZ) using EC and achieved complete MNZ degradation under operating conditions of a pH of 8.2, a CD of 6.0 mA/cm^2, and a treatment period of 14.6 minutes [31]. Furthermore, Arsand et al. (2013) explored the EC process for treating hospital effluent and observed a 30% increase in degradation efficiency by reducing the inter-electrode distance and increasing CD without pH adjustment. They also achieved the complete elimination of colloids and organic load from hospital wastewater [32]. Ahmadzadeh et al. (2017) also conducted an EC experiment to treat hospital effluent by eliminating ciprofloxacin and achieved 88.57% degradation efficiency at a pH of 7.78 and a CD of 12.5 mA/cm^2 using Al electrodes for 20 minutes [33].

6.4 SLUDGE OBTAINED FROM INDUSTRIAL WASTEWATER TREATMENT PLANTS

Sludge obtained from industrial wastewater treatment plants is a by-product of the treatment process that plays a crucial role in waste management and environmental protection. This sludge, also known as industrial or wastewater sludge, consists of solid particles, organic matter, inorganic compounds, and microorganisms that are separated during wastewater treatment. Understanding the composition, characteristics, and management of this sludge is essential for sustainable wastewater treatment and resource recovery practices.

The composition of industrial wastewater sludge can vary widely depending on the industry, wastewater characteristics, and treatment processes employed. The characteristics of industrial wastewater sludge also influence its handling, disposal, and potential reuse or recycling. Sludge can have different physical properties like moisture content, density, particle size distribution, and organic content, which affect its dewatering, transport, and storage requirements. High moisture content can make sludge bulky and costly to handle, while organic-rich sludge may be suitable for composting or anaerobic digestion to produce biogas.

6.4.1 Types and Composition of Sludge

Sludge refers to any semi-solid residue (typically water-based) generated as a by-product of a chemical or industrial process. It commonly arises from residual impurities in the initial material. Sludge is commonly encountered in sewage treatment plants, water purification facilities, refineries, manufacturing plants, or storage containers. Similarly, industrial sludge typically develops through the settling and sedimentation of biological or heavy metal impurities and waste from various processes. Its water content can vary depending on the original process that produced it, and there are multiple stages involved in its formation, treatment, and subsequent reprocessing. It can pose a risk to human health because the sludge and washings include a variety of chemical compounds.

The chemical composition of the sludge is primarily influenced by various factors such as (a) the types and amounts of chemicals present in the treated water, (b) the type and quantities of reagents utilized in the treatment procedure, (c) the type and amount of contaminants removed, and the type and concentration of the intermediates and by-products produced from the reactions taking place throughout the treatment process, all have an impact on the cleaning results [34]. All the materials present in groundwater and surface water are extracted from the water, accumulated and condensed into the sludge during the treatment process. Heavy metals including Cr, Zn, Pb, Sb, and Hg are present in higher concentrations in the sludge that is left over after the treatment of surface waters [35]. Consequently, for drinking water to meet safety regulations, it must be free of pollutants that are both naturally occurring and human-made. The sludge that is produced during the coagulation, ion exchange, and decarbonization processes usually contains the most harmful contaminants. Owing to its size, iron or aluminum hydroxide sediments, contaminants from treated water (bacteria, algae, protozoa, and minerals), and chemicals introduced during the treatment process, this sludge presents a serious ecological risk. Different kinds of sludge are produced throughout the treatment procedures, and they are discussed below.

6.4.1.1 Primary Sludge

The end product of the initial wastewater treatment process is the primary sludge. It is composed of solid contaminants that do not dissolve and form during pretreatment. Another name for the sediment that builds up at the bottom of the primary clarifier is *primary sludge*. The composition of this sediment changes depending on the features of the drainage area, but it usually contains a lot of organic debris, including paper, fibers, fruits, vegetables, and feces.

Primary sludge exhibits a thick, viscous consistency and has a water content of 93% to 97%. It is created during the initial treatment process when organic components and suspended solids are trapped by gravity sedimentation, which is frequently assisted by a primary clarifier.

6.4.1.2 Secondary Sludge

The secondary treatment method uses microorganisms to biologically break down the organic materials existing in the wastewater.

(a) **Activated sludge:** Nutrients and dissolving substances are taken out of the wastewater during the biological treatment step. This process uses a variety of oxygen-dependent bacteria and microorganisms to break down and eat the organic materials. This process produces sludge which is known as *activated sludge*, and it usually contains both dead and living biomass in addition to adsorbed, stored, organic, and mineral components.

(b) **Return activated sludge:** The activated sludge moves from the biological aeration basin into the final clarifier. The treated wastewater can be separated from the activated sludge flakes that settle at the bottom of the treatment unit. Return activated sludge is the part of the separated sludge which is recycled back into the aeration basin. Any excess secondary sludge is treated further and disposed of. Unused biomass must be taken out of the biological or secondary treatment system as extra sludge and disposed of to maintain a constant sludge age. This surplus sludge consists of biomass and particle debris that is not hydrolyzable by metabolism.

6.4.1.3 Digestive Sludge

The by-product of anaerobic digestion is *digested sludge*, which has an earthy smell and a black color. The organic solid content of anaerobic sludge usually varies from 45% to 60% depending on the degree of stabilization attained.

6.4.1.4 Tertiary Sludge

Tertiary sludge is generated through additional wastewater treatment processes involving the introduction of flocculants [36].

Table 6.4 displays the chemical analysis results of a typical sludge sample produced in steel, textile, pharmaceutical, tannery, and paper and pulp industries. The steel industry sludge sample indicates that Fe content ranges from 61% to 64.1% and CaO content ranges from 9% to 10%. Based on the chemical analysis, it is evident that Fe is primarily in its divalent state due to reduction during operational procedures. With its high iron content and significant CaO content, this sample is a great source of raw material for recycling in the sector of iron and steel. Furthermore, its lower Al_2O_3 content compared to iron ore presents an added advantage for reuse in steel plants [37]. Likewise, the textile effluent sludge (TES) sample exhibits a higher amount of SiO_2, approximately 57.829%. The presence of SiO_2 in TES indicates its strong binding capability. The lime content in the sample ranges from 16% to 17%. Lime influences concrete positively by enhancing uniformity, speeding up setting time, improving soundness, and maximizing compressive and flexural strength when used appropriately. Moreover, TES's substantial amounts of CaO and Fe_2O_3 further support its potential as a building material. Therefore, it can be inferred that TES can contribute to building strength to a significant extent [38]. Similarly, pharmaceutical industry sludge sample contains CaO and SiO_2 in the amount of 41.03% and 26.44%, respectively [39]. In addition, tannery industry sludge had a significant concentration of Cr(III) (14.1%) and hazardous Cr(VI) was completely absent. Similarly, a high amount of 14.8% was noted for Ca. However, the organic matter content was similarly significant (22%), perhaps due to the nature of the source material (leather). It is noteworthy to notice that the percentage of Mg, Na, Al, and Fe (<2.5%) in the sludge was significantly lower than that of the prior metals [40]. Furthermore, paper and pulp industry sludge exhibits a significant presence of CaO, which can be attributed to the inclusion of calcite in the raw fibrous material and the chemical additives used throughout the paper manufacturing process, including dyeing and uniformity enhancement. Elements with elevated CaO levels facilitate the formation of compounds upon contact with certain construction materials like cement, thereby enhancing mechanical strength [41].

6.4.2 Separation of Sludge and Settling Characteristics

Figure 6.2 illustrates the settling behavior of the EC-generated industrial sludge produced from the treatment of industrial wastewater using the EC process. The EC—treated liquids underwent additional settling processes. Initially, during the sludge settling test, a brief phase of relatively ineffective settling typically occurs, mainly attributed to Brownian motion; this phase is known as the *coalescence phase*. During this phase, the boundary between the flocs and the supernatant is not distinctly visible. Instead, there exists a *transitional region* between the sediment and the supernatant, where flocs of varying sizes settle independently. This phenomenon becomes more evident during EC using Al electrodes, where polymeric hydroxides create a gel-like substance. Following this initial phase, there is a consistent fall in the height of the solid–liquid interface, known as the *settling zone*, with a constant settling velocity. Subsequently, the sludge settling transitions into a phase of decreasing settling velocity known as the *transition settling* period. Lastly, *compression settling* occurs, and it is characterized by a very gradual change in the interface height.

TABLE 6.4
Composition of sludge generated from various industrial wastewater treatment plants.

Steel industry [37]		Textile industry [38]		Pharmaceutical industry [39]		Tannery industry [40]		Paper and pulp industry [41]	
Element	Inorganic component (wt%)	Element	Inorganic component (wt%)	Element	Inorganic component (wt%)	Element	Inorganic component (wt%)	Element	Inorganic component (wt.%)
Fe (total)	64.12	CaO	16.719	CaO	41.03	Cr (III)	14.1	SiO_2	8.03
FeO	79.58	Fe_2O_3	11.059	Fe_2O_3	1.10	Cr (VI)	---	Al_2O_3	9.46
Fe_2O_3	2.79	SiO_2	57.829	P_2O_5	7.72	Ca	14.8	Fe_2O_3	0.65
CaO	8.9	SO_3	---	SiO_2	26.44	Mg	2.4	K_2O	1.25
MgO	0.38	Al_2O_3	10.843	Al_2O_3	5.13	Na	1.5	TiO_2	0.40
SiO_2	0.71			K_2O	2.55	Al	0.5	SO_3	0.86
Al_2O_3	0.32			Na_2O	1.28	Fe	0.46	CaO	70.84
P	0.101			MgO	0.13	OM	22	C	16.8
MnO	0.10								

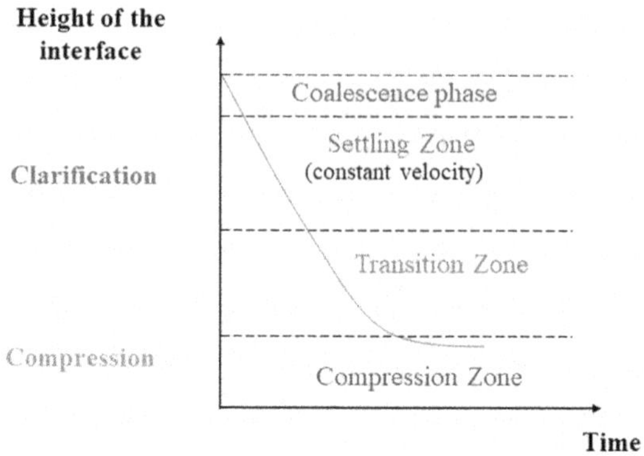

FIGURE 6.2 Sludge settling phases.

The resulting clear treated water was assessed for color, COD, and TS removal, and it was found that they showed a remarkable removal efficiency of 99%, 93%, and 89%, respectively [42].

6.5 THEORETICAL AND EXPERIMENTAL HYDROGEN PRODUCTION YIELDS DURING ELECTROCOAGULATION

Numerous research studies have highlighted the capabilities of the EC process in treating diverse types of wastewater, such as eliminating TSS, heavy metals, and dyes, breaking oil emulsions in water, addressing complex organic pollutants, and eliminating bacteria, viruses, and cysts. Despite the considerable research on EC for wastewater treatment, its large-scale implementation has been limited. One potential factor is the energy requirements for the EC process. According to previous studies, there has been no prior study on the extraction of hydrogen (H_2) from the EC process. Hydrogen is obtained as a significant by-product during the EC process, produced at the cathodes by water electrolysis. Through an efficient gas-liquid-solid separation technique, it is possible to extract high-quality H_2 from the EC process, which can both serve as a valuable energy source and reactant in various industrial routes.

According to Phalakornkule et al. (2010), the discharge from an EC reactor was continuously directed to the lower section of the gas separation column, initially containing water. As the effluent ascended to the top section of the column, gas bubbles ascended alongside it, carrying the suspended sludge particles with them. The gas exited through the gas outlet positioned at the column's upper section and water exited through a separate water discharge opening near the top of the column (refer to Figure 6.3). The gas is accumulated in a liquid displacement gas meter. When liquid flow rates are high, light and dispersed sludge are carried away with the liquid stream, resulting in minimal solid accumulation within the gas separation column.

Table 6.5 displays the amount of H_2 obtained using the EC method. This amount was determined by multiplying the gas volume measured using the liquid displacement gas meter by the H_2 purity assessed using GC. Initially, the gas mixture

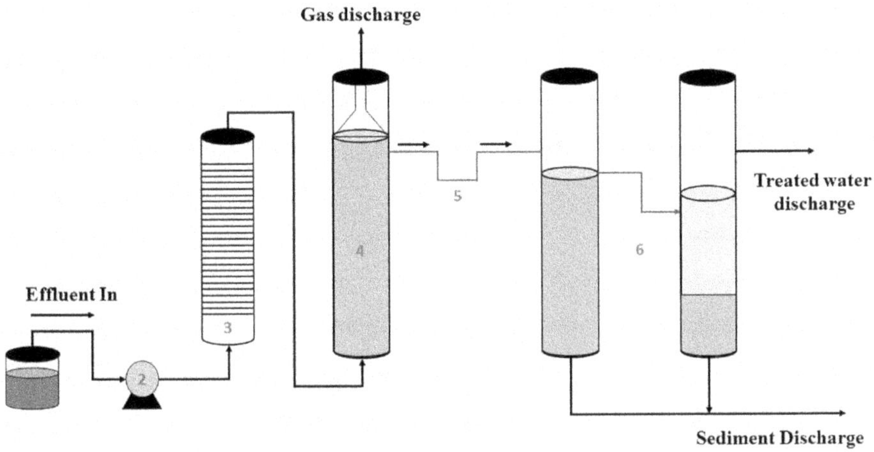

FIGURE 6.3 EC treatment setup: (1) effluent storage tank, (2) pump, (3) EC reactor, (4) gas separating column, (5) U-tube type connector, and (6) two sedimentation columns [43].

TABLE 6.5
Comparitive analysis of theoretical energy yield from the harvested H_2 product and energy requirement for the electrocoagulation process (Reproduced with permission from Phalakornkule et al. (2010) @ Elsevier [43]).

Current density (J, A/m^2)	Electrocoagulation time (t, minutes)	Theoretical amount of H_2: Mole	Theoretical amount of H_2: (x 10^3 m^3 at the normal condition)[a]	Experimental H_2 volume (x 10^3 m^3 at the normal condition)	H_2 yield (experimental/ theoretical)
30	5	0.0244	0.587	0.521	0.89
40	5	0.0325	0.783	0.733	0.94
30	10	0.0488	1.173	1.039	0.89
40	10	0.0651	1.565	1.550	0.99

[a] Assuming that the ideal gas law under standard conditions as PV = nRT, where T equals 293 K and P equals 1.01325 × 10^5 Pa.

comprised air and H_2. As the operational time progressed, the H_2 purity increased due to the displacement of air by the produced H_2.

The hydrogen yield was determined by dividing the amount of hydrogen obtained through experimentation by the amount of H_2 that was theoretically calculated using Equation (6.5). In the EC process, DC supplied to the electrodes initiates oxidation and reduction reactions. Equation (6.4) describes the relationship between the amount of H_2 generated and the supplied current.

$$n_{H_2} = \frac{I.t}{F}.H \tag{6.4}$$

Likewise, Equation (6.5) represents the relationship between the amount of hydrogen generated, the effective electrode area, and the supplied current density.

$$n_{H_2} = \frac{J.A.t}{F}.H \tag{6.5}$$

where n_{H_2} represents the quantity of H_2 generated (moles), I denotes the supplied current (A), J represents the supplied current density (A/m²), A is the effective area of the electrode (m²), t signifies the EC treatment time (s), F is Faraday's constant (96,500 C/mole of electrons), and H signifies the number of H_2 molecules generated for each electron used in the redox reactions. The value of H remains the same regardless of the anode used, whether it is Fe or Al. In both cases, H is equal to 1/2.

The H_2 yield ranged from 0.89 to 0.94 for the first three sets of CD and EC treatment time (CD30T5, CD40T5, and CD30T10). This suggests that around 6 to 11% of produced H_2 was released from the gas separation column along with the liquid effluent stream. The process was running continuously when the measurements were taken, and the majority of air had been replaced by the H_2 gas in the system. The yield measurement was carried out in the first three cases when the H_2 purity was noticeably high (>85%). At such high purity levels, even a minor gas product loss would lead to a significant decrease in the H_2 output. The hydrogen purity in the fourth scenario (CD40T10) was about 35% since the measurement was made during the system's beginning phase when there was still a significant amount of air present. If the amount of gas exiting the column with the liquid effluent is equal to that during the first three scenarios, then there would be less H_2 loss, which could result in a higher yield [43].

6.6 CHALLENGES AND FUTURE PERSPECTIVES

Despite the long-term use of electrocoagulation (EC), there is a lack of a systematic approach in the existing literature for designing and operating EC reactors. While EC has been proved to be effective in treating water and wastewater on a laboratory scale, transitioning to industrial-scale applications poses challenges. This is primarily because current research primarily concentrates on removing specific pollutants from wastewater through batch experiments. A significant challenge facing this technology is its high electricity consumption during wastewater treatment, which directly impacts operational expenses. Consequently, the success of this treatment process hinges on its capability to meet commercial standards by reducing operational and maintenance costs and generating returns on investment swiftly. Additionally, the technology faces stiff competition from established water treatment methods like membrane separation, adsorption, and biotechnology. Integrating the EC technique with these existing methods could enhance its prospects for success [44].

6.7 CONCLUSION

The EC process is known for its high processing efficiency, ease of operation, and minimal sludge generation and emerges as a promising and efficient technology for treating industrial effluents. It proves effective in the removal of various pollutants such as suspended solids, organic compounds, heavy metals, and inorganic anions. The effectiveness of EC relies on various operational parameters, including the electrode type, applied current density, electrolysis duration, electrolyte support, pH levels, and power supply type. This chapter comprehensively covers the removal of contaminants from various industrial wastewater, along with the comparative study of hydrogen gas generated as a by-product and theoretical hydrogen. However, it is important to note that despite its potential, EC encounters challenges in the real-life industrial effluent treatment system, particularly regarding the lack of systematic design and operation guidelines for EC reactors and their high electricity consumption. These challenges require further attention and research efforts in future.

REFERENCES

[1] M. Bharti, P.P. Das, M.K. Purkait, A review on the treatment of water and wastewater by electrocoagulation process: Advances and emerging applications, *J. Environ. Chem. Eng*. 11 (2023) 111558. https://doi.org/10.1016/j.jece.2023.111558.

[2] P.P. Das, M. Sharma, M.K. Purkait, Recent progress on electrocoagulation process for wastewater treatment: A review, *Sep. Purif. Technol.* 292 (2022) 121058. https://doi.org/10.1016/j.seppur.2022.121058.

[3] S. Dhara, A.D. Sontakke, N. Shekhar Samanta, R. Venkata Sai Uppaluri, M.K. Purkait, Biowaste-alkaline lignin and GO integrated polysulfone ultrafiltration membrane fabrication for Pb2+ and Eosin Y dye removal, *Sep. Purif. Technol.* 341 (2024) 126894. https://doi.org/10.1016/j.seppur.2024.126894.

[4] N. Shekhar Samanta, P.P. Das, S. Dhara, M.K. Purkait, An overview of precious metal recovery from steel industry slag: Recovery strategy and utilization, *Ind. Eng. Chem. Res.* 62 (2023) 9006–9031. https://doi.org/10.1021/acs.iecr.3c00604.

[5] P.P. Das, S. Dhara, M.K. Purkait, Chapter 13—Ozone-based oxidation processes for the removal of pharmaceutical products from wastewater, in: M.P. Shah, P. Ghosh (Eds.), *Development in Wastewater Treatment Research and Processes*, Elsevier, 2024: pp. 287–308. https://doi.org/10.1016/B978-0-443-19207-4.00003-3.

[6] J.-w. Feng, Y.-b. Sun, Z. Zheng, J.-b. Zhang, S. Li, Y.-c. Tian, Treatment of tannery wastewater by electrocoagulation, *J. Environ. Sci.* 19 (2007) 1409–1415. https://doi.org/10.1016/S1001-0742(07)60230-7.

[7] A. Shahedi, A.K. Darban, F. Taghipour, A. Jamshidi-Zanjani, A review on industrial wastewater treatment via electrocoagulation processes, *Curr. Opin. Electrochem.* 22 (2020) 154–169. https://doi.org/10.1016/j.coelec.2020.05.009.

[8] J. Ahmed, A. Thakur, A. Goyal, Industrial wastewater and its toxic effects, *Biol. Treat. Ind. Wastewater* (2021) 1–14. https://doi.org/10.1039/9781839165399-00001.

[9] S. Mohan, N. Muralimohan, K. Vidhya, C.T. Sivakumar, A case study on-textile industrial process, characterization and impacts of textile effluent, *Indian J. Sci. Res.* 17 (2017) 80–85.

[10] R.S. Rana, P. Singh, V. Kandari, R. Singh, R. Dobhal, S. Gupta, A review on characterization and bioremediation of pharmaceutical industries' wastewater: An Indian perspective, *Appl. Water Sci.* 7 (2017) 1–12. https://doi.org/10.1007/s13201-014-0225-3.

[11] A. Rawat, A. Srivastava, A. Bhatnagar, A.K. Gupta, Technological advancements for the treatment of steel industry wastewater: Effluent management and sustainable treatment strategies, *J. Clean. Prod.* 383 (2023) 135382. https://doi.org/10.1016/j.jclepro.2022.135382.
[12] M.A. Moktadir, M. Maliha, F. Tujjohra, S.A. Munmun, M.S. Alam, M.A. Islam, M.M. Rahman, Treatment of tannery wastewater by different membrane bioreactors: A critical review, *Environ. Adv.* 15 (2024) 100478. https://doi.org/10.1016/j.envadv.2023.100478.
[13] O. Ashrafi, L. Yerushalmi, F. Haghighat, Wastewater treatment in the pulp-and-paper industry: A review of treatment processes and the associated greenhouse gas emission, *J. Environ. Manage.* 158 (2015) 146–157. https://doi.org/10.1016/j.jenvman.2015.05.010.
[14] S. Sandhya, K. Sarayu, K. Swaminathan, Determination of kinetic constants of hybrid textile wastewater treatment system, *Bioresour. Technol.* 99 (2008) 5793–5797. https://doi.org/10.1016/j.biortech.2007.10.011.
[15] M. Kobya, O.T. Can, M. Bayramoglu, Treatment of textile wastewaters by electrocoagulation using iron and aluminum electrodes, *J. Hazard. Mater.* 100 (2003) 163–178. https://doi.org/10.1016/S0304-3894(03)00102-X.
[16] M. Kobya, E. Gengec, M.T. Sensoy, E. Demirbas, Treatment of textile dyeing wastewater by electrocoagulation using Fe and Al electrodes: Optimisation of operating parameters using central composite design, *Color. Technol.* 130 (2014) 226–235. https://doi.org/10.1111/cote.12090.
[17] U. Tezcan Un, E. Aytac, Electrocoagulation in a packed bed reactor-complete treatment of color and cod from real textile wastewater, *J. Environ. Manage.* 123 (2013) 113–119. https://doi.org/10.1016/j.jenvman.2013.03.016.
[18] M. Kobya, E. Gengec, E. Demirbas, Operating parameters and costs assessments of a real dyehouse wastewater effluent treated by a continuous electrocoagulation process, *Chem. Eng. Process. Process Intensif.* 101 (2016) 87–100. https://doi.org/10.1016/j.cep.2015.11.012.
[19] A. Benhadji, M. Taleb Ahmed, R. Maachi, Electrocoagulation and effect of cathode materials on the removal of pollutants from tannery wastewater of Rouïba, *Desalination* 277 (2011) 128–134. https://doi.org/10.1016/j.desal.2011.04.014.
[20] I.A. Şengil, S. Kulaç, M. Özacar, Treatment of tannery liming drum wastewater by electrocoagulation, *J. Hazard. Mater.* 167 (2009) 940–946. https://doi.org/10.1016/j.jhazmat.2009.01.099.
[21] R.N. Bharagava, S. Mishra, Hexavalent chromium reduction potential of Cellulosimicrobium sp. isolated from common effluent treatment plant of tannery industries, *Ecotoxicol. Environ. Saf.* 147 (2018) 102–109. https://doi.org/10.1016/j.ecoenv.2017.08.040.
[22] F.R. Espinoza-Quiñones, M.M.T. Fornari, A.N. Módenes, S.M. Palácio, F.G. da Silva, N. Szymanski, A.D. Kroumov, D.E.G. Trigueros, Pollutant removal from tannery effluent by electrocoagulation, *Chem. Eng. J.* 151 (2009) 59–65. https://doi.org/10.1016/j.cej.2009.01.043.
[23] S. Elabbas, N. Ouazzani, L. Mandi, F. Berrekhis, M. Perdicakis, S. Pontvianne, M.N. Pons, F. Lapicque, J.P. Leclerc, Treatment of highly concentrated tannery wastewater using electrocoagulation: Influence of the quality of aluminium used for the electrode, *J. Hazard. Mater.* 319 (2016) 69–77. https://doi.org/10.1016/j.jhazmat.2015.12.067.
[24] D. Syam Babu, T.S. Anantha Singh, P. V. Nidheesh, M. Suresh Kumar, Industrial wastewater treatment by electrocoagulation process, *Sep. Sci. Technol.* 55 (2020) 3195–3227. https://doi.org/10.1080/01496395.2019.1671866.
[25] R. Sridhar, V. Sivakumar, V. Prince Immanuel, J. Prakash Maran, Treatment of pulp and paper industry bleaching effluent by electrocoagulant process, *J. Hazard. Mater.* 186 (2011) 1495–1502. https://doi.org/10.1016/j.jhazmat.2010.12.028.
[26] M. Zaied, N. Bellakhal, Electrocoagulation treatment of black liquor from paper industry, *J. Hazard. Mater.* 163 (2009) 995–1000. https://doi.org/10.1016/j.jhazmat.2008.07.115.
[27] M. Uğurlu, A. Gürses, Ç. Doğar, M. Yalçin, The removal of lignin and phenol from paper mill effluents by electrocoagulation, *J. Environ. Manage.* 87 (2008) 420–428. https://doi.org/10.1016/j.jenvman.2007.01.007.

[28] S. Mahesh, B. Prasad, I.D. Mall, I.M. Mishra, Electrochemical degradation of pulp and paper mill wastewater. Part 2. Characterization and analysis of sludge, *Ind. Eng. Chem. Res.* 45 (2006) 5766–5774. https://doi.org/10.1021/ie0603969.
[29] S. Singh, S. Singh, S.L. Lo, N. Kumar, Electrochemical treatment of Ayurveda pharmaceuticals wastewater: Optimization and characterization of sludge residue, *J. Taiwan Inst. Chem. Eng.* 67 (2016) 385–396. https://doi.org/10.1016/j.jtice.2016.08.028.
[30] S. Veli, A. Arslan, D. Bingöl, Application of response surface methodology to electrocoagulation treatment of hospital wastewater, *Clean Soil Air Water* 44 (2016) 1516–1522. https://doi.org/10.1002/clen.201500729.
[31] S. Ahmadzadeh, M. Dolatabadi, Electrochemical treatment of pharmaceutical wastewater through electrosynthesis of iron hydroxides for practical removal of metronidazole, *Chemosphere* 212 (2018) 533–539. https://doi.org/10.1016/j.chemosphere.2018.08.107.
[32] D.R. Arsand, K. Kümmerer, A.F. Martins, Removal of dexamethasone from aqueous solution and hospital wastewater by electrocoagulation, *Sci. Total Environ.* 443 (2013) 351–357. https://doi.org/10.1016/j.scitotenv.2012.10.100.
[33] S. Ahmadzadeh, A. Asadipour, M. Pournamdari, B. Behnam, H. Reza, M. Dolatabadi, Removal of ciprofloxacin from hospital wastewater using electrocoagulation technique by aluminum electrode : Optimization and modelling through response surface methodology, *Process Saf. Environ. Prot.* 109 (2017) 538–547. https://doi.org/10.1016/j.psep.2017.04.026.
[34] X. Wang, M. Li, J. Liu, J. Qu, Occurrence, distribution, and potential influencing factors of sewage sludge components derived from nine full-scale wastewater treatment plants of Beijing, China, *J. Environ. Sci.* 45 (2015) 233–239. https://doi.org/10.1016/j.jes.2016.01.005.
[35] H. Geng, Y. Xu, L. Zheng, H. Gong, L. Dai, X. Dai, An overview of removing heavy metals from sewage sludge: Achievements and perspectives, *Environ. Pollut.* 266 (2020) 115375. https://doi.org/10.1016/j.envpol.2020.115375.
[36] A. Qarani, M. Sabah, Wastewater sludge characteristics, treatment techniques and energy production, *Reciklaza i Odrziv. Razvoj* 15 (2022) 9–26. https://doi.org/10.5937/ror2201009a.
[37] B. Das, S. Prakash, P.S.R. Reddy, V.N. Misra, An overview of utilization of slag and sludge from steel industries, *Resour. Conserv. Recycl.* 50 (2007) 40–57. https://doi.org/10.1016/j.resconrec.2006.05.008.
[38] U.S. Patil, S.P. Raut, Utilization of textile effluent sludge to create concrete, *Mater. Today Proc.* (2023). https://doi.org/10.1016/j.matpr.2023.07.095.
[39] Y. Zhou, F. Lin, Z. Ling, M. Zhan, G. Zhang, D. Yuan, Comparative study by microwave pyrolysis and conventional pyrolysis of pharmaceutical sludge: Resourceful disposal and antibiotic adsorption, *J. Hazard. Mater.* 468 (2024) 133867. https://doi.org/10.1016/j.jhazmat.2024.133867.
[40] E. Kokkinos, A. Zouboulis, A step by step investigation of Cr(III) recovery from tannery waste, *Proceedings* 48 (2020) 6436. https://doi.org/10.3390/ecws-4-06436.
[41] A.R.G. de Azevedo, J. Alexandre, L.S.P. Pessanha, R. da S.T. Manhães, J. de Brito, M.T. Marvila, Characterizing the paper industry sludge for environmentally-safe disposal, *Waste Manag.* 95 (2019) 43–52. https://doi.org/10.1016/j.wasman.2019.06.001.
[42] A.A. Amusa, M.R. Taib, W.Z. Xian, Continuous flow electrochemical process for sanitary landfill leachate treatment: Role of inlet flow rate and current density, *Water Air Soil Pollut.* 234 (2023). https://doi.org/10.1007/s11270-023-06509-z.
[43] C. Phalakornkule, P. Sukkasem, C. Mutchimsattha, Hydrogen recovery from the electrocoagulation treatment of dye-containing wastewater, *Int. J. Hydrogen Energy* 35 (2010) 10934–10943. https://doi.org/10.1016/j.ijhydene.2010.06.100.
[44] D.T. Moussa, M.H. El-Naas, M. Nasser, M.J. Al-Marri, A comprehensive review of electrocoagulation for water treatment: Potentials and challenges, *J. Environ. Manage.* 186 (2017) 24–41. https://doi.org/10.1016/j.jenvman.2016.10.032.

7 Advancements in a Hybrid Electrocoagulation Process

7.1 INTRODUCTION

Water scarcity is one of the major challenges of the twenty-first century. An estimated 7.8 million people worldwide pass away each year as a result of inadequate sanitation and water hygiene, and almost 2 billion people lack easy access to safe drinking water [1]. Furthermore, every year, drought affects about 55 million people, resulting in economic losses of approximately US$5 billion. Water use has risen by 1% annually since 1980, and this trend is predicted to continue until 2050. Water treatment procedures can generally be divided into two categories: separation techniques and degradation techniques [2, 3]. Using a separation technique, pollutants in water are moved from one phase to another, usually into the solid phase. The adsorption process is the best example of this phenomenon. Nevertheless, extra treatment is necessary since separation techniques, such as membrane separation, concentrate the pollutants in the water phase. Concurrently, the degrading approach (used to remove organic pollutants) frequently causes the impurities to break down until full mineralization is attained [4]. Both biological [5] and physicochemical [6] processes, as well as more recent methods like adsorption [7] and membrane filtration [8], are included in conventional wastewater treatment. Strict regulations, a longer retention period, a greater footprint, and the production of unwanted by-products are necessary for biological processes [9]. Chemical procedures necessitate the significant use of chemicals, which raises the process's overall cost and complicates downstream activities, raising the possibility of secondary pollution. Furthermore, unless they are paired with thorough pretreatment procedures, adsorption and membrane filtration by themselves are insufficient for treating highly contaminated wastewater. Adsorbents gradually lose their potential as the number of cycles grows; the adsorption process is highly dependent on the pH of the solution and necessitates a longer treatment period [3, 10]. Additionally, the renewal of adsorbents requires the use of vacuum or steam. Furthermore, as pollutants clog pores and cause flux loss, membrane filtration loses efficiency over time. As a result, electrochemical techniques are heavily the focus of contemporary wastewater treatment research [11, 12].

Because the electrocoagulation (EC) technique is versatile, simple to use, and capable of handling a wide range of contaminants, it has been claimed to be an

DOI: 10.1201/9781003499671-7

effective method for treating both water and wastewater. For the treatment of highly contaminated wastewater, electrocoagulation is seen to be a feasible substitute for chemical coagulation because of its high removal rate and lack of additional chemicals. Unlike chemical coagulation, which involves the addition of a coagulant, electrocoagulation does not carry the risk of producing secondary contaminants [13]. Additionally, during the EC process, gas bubbles occur, which helps pollutants float on the water surface. Furthermore, the flocs generated by electrocoagulation have a constant, big particle size that is easily separated by decantation or straightforward filtration. Because the flocs are composed of an acid-resistant chemical rather than water, they can be removed more easily by filtration. In comparison to chemical coagulation, less sludge is formed during electrocoagulation since the majority of the flocs are agitated. This shows that when an electric field is applied to aid in the coagulation and flotation of impurities, electrocoagulation is highly effective in removing even the smallest colloidal particles [14]. Therefore, a variety of contaminants, including total suspended particles, total organic carbon, oil emulsions, heavy metals, chemical oxygen demand, color, and turbidity, can be eliminated from both water and wastewater using the EC method. Numerous studies have thoroughly explored the benefits of using the EC technique for wastewater treatment. But a significant problem with the EC procedure is the removal of sludge from the treated solution, necessitating a second post-treatment step [15, 16]. Therefore, to increase the effectiveness of the EC approach and address issues with environmental applications, a great deal of research needs to be done on the coupling of EC with other existing treatment techniques. As far as we are aware, no research has been done on the whole set of freestanding and hybrid electrocoagulation systems as well as the various advancements and uses of these systems in the treatment of wastewater and water. The objective here is to couple the EC method to successfully apply it to multiple applications, such as wastewater remediation, sustainable water desalination, and water reuse. In the past ten years, there has been a surge in interest in hybrid electrocoagulation procedures among researchers. Carefully selecting how to hybridize the electrocoagulation process with other treatment approaches can reduce energy consumption and increase removal efficiency [3].

This chapter elaborately assesses the performance efficacy of membrane filtration (microfiltration/ultrafiltration/nanofiltration/reverse osmosis), oxidation (ozonation/hydrogen peroxide/ultrasound/ultraviolet radiation), adsorption (activated carbon/biofiltration), and other integrated processes when combined with electrocoagulation. In the current review paper, these hybrid approaches are covered in great detail, with an emphasis on energy consumption and wastewater treatment, as well as a number of issues and suggestions for the future. In order to gain a crucial awareness of both the single and combined methods, this research also looks at the recent developments in the scaling up of the combined EC method.

7.2 INTEGRATED ELECTROCOAGULATION AND MEMBRANE PROCESSES

7.2.1 Electrocoagulation—Microfiltration (EC-MF)

Microfiltration (MF) is a method of filtering particles with pore diameters of at least 0.1 μm. Because of their larger pore size, MF membranes foul more easily than other membrane types. In order to reduce fouling on the ceramic MF membrane, a higher crossflow velocity (5–7 m/s) is produced at the retentate side by increasing the retentate recirculation pump rate. High wall shear stress is produced as a result, and the foulants are removed from the membrane surface [2, 4, 17]. However, electrocoagulation is a practical and affordable method that may be used as a pretreatment procedure to get rid of organic pollutants and suspended particles, which helps to keep membrane fouling at bay. The pH of the solution, the anode material, the length of the treatment, and the kind and concentration of pollutants all have a significant impact on how well EC filters. The use of a hybrid EC-MF approach as an alternative treatment to UF for the removal of natural organic matter (NOM) demonstrated a number of benefits such as the ability to sustain high permeate flux, low fouling, and high removal rates [18, 19]. A batch EC-MF technique was implemented by Changmai et al. (2019) to treat oily wastewater emulsions that contain grease, oil, and heavy metals [20]. In order to investigate the impact on the effectiveness of pollutant removal, various operational parameters were changed during the pretreatment process, including electrode gap, starting pH, and current density. It was found that a significant quantity of flocs was formed when a voltage differential was applied across the electrode surface. The primary functions of these flocs are to remove undesirable metals from the solution and lower the amounts of oil and grease from 35 to 10.2 mg/L. The flocculants were then eliminated from the treated water and the pollutant content was further decreased by applying the MF approach. It was found that the permeate flow rose from 264 to 423 L/h m^2 at a pressure of 98 kPa as the sintering temperatures were raised from 700 to 1000°C. The hybrid EC-MF method was found to have further decreased the oil and grease content below each of their individual allowable limits (0.2 mg/L).

7.2.2 Electrocoagulation—Ultrafiltration (EC-UF)

In essence, ultrafiltration (UF) is a low pressure, economically viable membrane-based treatment method. Particle collection at the membrane's surface and across its pores is caused by process parameters such as membrane fouling, which significantly reduces the efficiency of water recovery [21, 22]. In the UF membrane separation process, suspended solids and organic contaminants are regarded as the essential foulants. Because of this, the EC-assisted UF filtration procedure could be a highly helpful strategy for lowering membrane fouling by lowering organic contaminants and suspended particles. Additionally, lowering the concentration of pollutants in the feed water for the UF process has the potential to greatly enhance the hybrid process's overall water quality [23]. Sardari et al. (2018) employed five Al electrodes with an inter-electrode spacing of 0.9 cm and a current density of 3 mA/cm^2 to remove TSS, BOD, and COD from poultry wastewater using EC as a pretreatment method [24]. The electrode arrangement was set up in the BP-S mode. 84% of TSS, 59% of COD, and 87% of BOD were removed efficiently after pretreatment using the EC procedure. It was discovered that the process used very little energy (0.15 kW h/m^3). The initial permeate flux rises during the UF process while the flux declination rate falls as a result of the previously indicated contaminants' earlier degradation. The regenerate cellulose (RC) separator used in the UF process had a flow rate of 1.2 L/min and an applied pressure of 70 kPa. High water permeate quality was achieved at the end of the hybrid (EC-UF) process, with removal efficiencies of 100%, 92%, and 98% for TSS, COD, and BOD, respectively. Nonetheless, the UF method was used before EC in a 2011 study by Yahiaoui et al. (2011) to remove COD in olive mill effluents [25]. The COD was successfully degraded by the UF membrane, resulting in a 96% removal efficiency and a decrease in the COD amount from 28 to 1.1 g dm^3. The surface water discharge standard was not met even after the UF procedure. As a result, an EC cell fitted with iron electrode was used to treat the permeate, which had an average COD value of 1.1 g dm^3. It was investigated how the COD removal effectiveness was affected by the operating parameters such as solution pH, current density, NaCl concentration, and surface electrode/reactor volume ratio. According to the reports, solution pH in the range of 4.5–6.5 and an ideal current density of 93.3 A/m^2 produced the best COD elimination effectiveness. Under ideal operating conditions, the EC procedure dramatically lowered the COD amount from 1.1 to 78 mg dm^3, allowing olive mill effluents to be released directly into the environment.

7.2.3 Electrocoagulation—Nanofiltration (EC-NF)

The membrane technique known as nanofiltration (NF) is pressure-driven and has pore sizes in between those of reverse osmosis (RO) and ultrafiltration (UF) membranes ranging between 0.2 and 2 nm. NF membranes can be used in conjunction with an appropriate pretreatment procedure to successfully prevent membrane fouling, especially in the case of wastewaters with high pollutant concentrations [26]. As a pretreatment measure, the EC method can be effectively applied to raise feed water quality and reduce membrane fouling. The EC process was chosen because it uses a few chemicals, effectively removes pollutants, and has minimal operating costs. Tavangar et al. (2019) used both independent and combined EC and NF methods to treat wastewater from the textile sector [27]. It has been claimed that the high color content and other contaminants that remain in the treated EC solution can be completely destroyed by the combined EC-NF process, which also enhances the process performance. Compared to single EC and NF, the decolorization rate was enhanced to about 2% and 8%, respectively. In addition, compared to the Fe and Ti electrodes, the flux is increased by 76% when EC is used with Al electrodes prior to the NF procedure. Furthermore, Gönder et al. (2020) used the EC-NF process to treat the carwash wastewater, which allowed for the efficient reuse of the treated wastewater as rinse water [28]. Water from car washes was pretreated using the EC procedure with the Fe electrode at varying temperatures, stirring rates, and electrode arrangement types. Under ideal conditions of 25°C, 250 rpm, and MP-P setup, EC performed best in terms of energy usage and pollutant removal. Desal 5DL and NF 270 membranes were used to repeat the NF process on the electrocoagulated samples. The carwash wastewater's oil and COD contents were lowered by 91% and 89%, respectively, and its energy consumption was estimated to be 2.7 kWh/m^3. Consequently, it was determined that wastewater recycling was greatly aided by the hybrid EC-NF method.

7.2.4 Electrocoagulation—Reverse Osmosis (EC-RO)

To increase process efficiency and avoid membrane fouling constraints, reverse osmosis (RO) is widely combined with other water treatment methods. The RO process is significantly impacted by membrane fouling, which includes both organic fouling and biofouling. The adhesion and buildup of bacteria, yeast, and fungi on the membrane surface results in biofouling. Thus, in order to prevent RO membrane fouling, a pretreatment method is needed [29, 30]. It has been demonstrated that applying the EC process before RO lowers the organic matter content, which causes organic and biofouling. In order to reduce membrane fouling limits, Zhao et al. (2014) used the EC method before RO during the treatment of synthetic water [31]. Iron anodes and graphite cathodes made up a total of six electrodes. An electrode spacing of 1.5 cm was maintained and the electrodes were coupled in an MP-P arrangement. Turbidity, COD, and hardness were found to have removal efficiencies of 93.8%, 66.5%, and 85.8%, respectively. The RO process was used to further treat the electrocoagulated samples, with an applied pressure of 2 MPa and a flow rate of 0.8 L/min. After 60 minutes of operation, there was a sudden drop in the water flux because the feed water included a high quantity of organic compounds and scaling ions. The water flux for pretreated and raw water dropped to 26 L/m^2 h and 4 L/m^2 h, respectively, after 90 minutes. The allowable filtering period was raised by 55% as a result of the pretreatment procedure using EC. Furthermore, the combined EC-RO method demonstrated exceptionally enhanced elimination rate for all contaminants, including hardness (99.5%), turbidity (99.1%), COD (95.5%), and oil (>99%). Similar outcomes were also reported by Sadeddin et al. (2011) while treating well water using a hybrid EC-RO technique [32]. Two stainless steel electrodes were coupled to a DC power source to form the EC reactor. It was discovered that the removal efficiency of turbidity and TSS were 98% and 99%, respectively. After being pretreated, the RO process was applied to the feed water, and it ran for six months. The RO plants are thought to benefit from the extremely low silt density index (3%/min) present in the input water. According to the study's findings, the RO system is much improved using EC as a pretreatment procedure. Additionally, the dissolved solids were successfully removed by the RO process, whereas the combined EC-RO method showed greater potential for the elimination of the target contaminants. However, Hakizimana et al. (2016) found that the poor hardness removal rate of EC made it inefficient in degrading scaling ions during the evaluation of EC as a pretreatment method for desalination of seawater [33]. The procedure required 0.17 kW h/m^3 of energy. Hardness was only removed up to 10%, despite the fact that the elimination efficacy of absorbance (82%) and DOC (58%) were determined to be adequate. The scaling ions were consequently improperly eliminated during the pretreatment procedure. With the exception of hardness, decreasing the solution pH and inter-electrode spacing boosted the rate of removal of other contaminants. In addition, the investigation found that aromatic chemicals degraded more quickly than aliphatic chemicals [2].

7.3 INTEGRATED ELECTROCOAGULATION—OXIDATION PROCESSES

7.3.1 Electrocoagulation—Ozonation (EC-O_3)

Ozonation (O_3) is a very efficient method that breaks down dyes and other wastewater contaminants using ozone, a potent oxidant. Ozone oxidation is a phenomenon that can be caused by an *in situ* hydroxyl radical attack that is indirectly created or by a direct ozone attack through electrophiles. The pace at which the contaminants oxidize is accelerated by increasing the O_3 dosage. O_3, on the other hand, is unable to eliminate contaminants with high molecular weights; for this reason, using EC as a pretreatment procedure can significantly increase process efficiency [34, 35]. Figure 7.1 displays the hybrid electrocoagulation-ozonation schematic diagram. The effectiveness of the hybrid EC-O_3 technique for removing high color concentration from saline water was determined by Bili´nska et al. (2019) [36]. The saline water's original pH value was recorded as 11.82. Eleven Al/Fe electrodes joined in a BP-S configuration made up the EC reactor. 85% of the color was removed after 10 minutes of electrolysis. However, because of the extremely high pH of the solution, the treatment period had no discernible impact on the rate of elimination. The electrocoagulated samples were treated further in a 1L semi-batch stirred reactor. Ozone was kept at a concentration and flow rate of 43 mg/L and 41 L/h, respectively. After 50 minutes of treatment, the O_3 procedure achieved a 92% color removal efficiency and a 56.3 kW h/m^3 energy usage. Nevertheless, after 10 minutes, the hybrid EC-O_3 method produced an overall

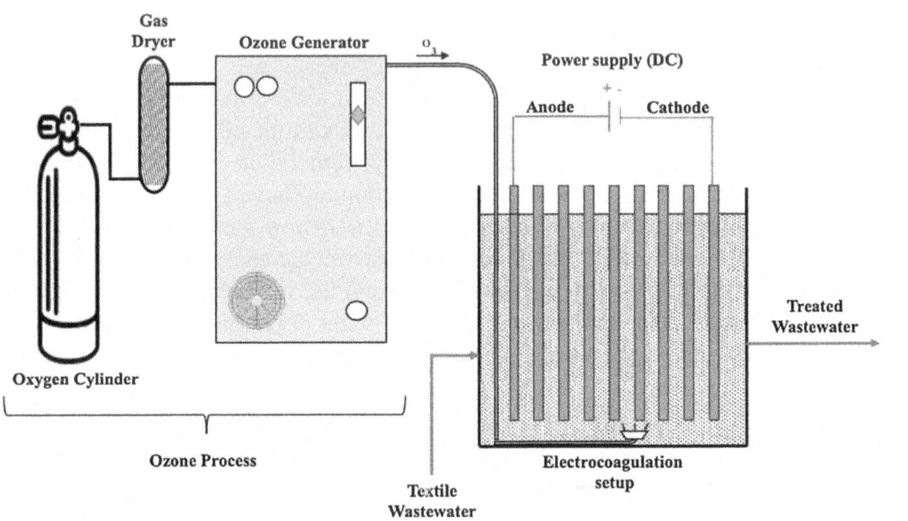

FIGURE 7.1 A schematic representation of the hybrid EC-O_3 method (Reproduced with permission from Almukdad et al. (2021) @ Elsevier [2]).

98% decolorization. According to the study's findings, applying the EC procedure before O_3 dramatically reduced the ozonation duration from 50 minutes to 10 minutes, cutting the ozonation length by 80% and conserving five times the dosage of ozone. Moreover, the hydroxyl radicals produced when hydrogen peroxide breaks down are catalyzed by the Al/Fe species produced by the EC process. As a result, there is an improvement in the generation of hydroxyl radicals in solution, which increases process efficiency [37]. Thus, the hybrid EC-O_3 method removes pollutants at a substantially faster pace than the O_3 process operating alone.

7.3.2 Peroxi-electrocoagulation (EC-H_2O_2)

Advanced oxidation processes (AOPs) based on electrochemistry are environmentally friendly methods of treating wastewater by breaking down dangerous and long-lasting pollutants. EAOPs have the capability to restrict and address the pollution problems resulting from the clean electron reagent when used in wastewater treatment. The advantages of the peroxi-EC technique are explained by the chemical processes that take place during the oxidation process. H_2O_2 and a ferrous ion source are essential for the oxidation process to occur. Sacrificial Fe anodes are used as the Fe^{2+} source in the peroxi-EC process, and external H_2O_2 addition is made. Furthermore, the ongoing regeneration of Fe^{2+} ions is contingent upon the decrease of Fe^{3+} species [38, 39]. The treatment of distillery effluents was carried out by Asaithambi et al. (2016) using the electrochemically induced oxidation technique [40]. The EC reactor was made up of two Fe electrodes coupled in a BP-S arrangement, spaced 3 cm apart. H_2O_2 was added to the solution at the beginning of the EC process, which increased the COD removal effectiveness of EC by 13%. In addition, the process's energy usage was 15% lower. Overall, distillery effluents were completely decolored, COD was reduced by 86%, and 1.1 kW h/m^3 of energy was used when H_2O_2 was added to the EC process. Yüksel et al. (2009) also looked into the effectiveness of the peroxi-EC technique for treating wastewater contaminated with sodium dodecyl sulfate (SDS) [41]. Inside the EC cell, a total of four Fe electrodes were arranged vertically with 2.5 cm apart. The BP-S connection mode was used to configure the electrodes. The SDS removal effectiveness of the hybrid method was 82.3%. Furthermore, 1.63 kW h/kg SDS was found to be the average energy usage. It was determined that, in comparison to individual H_2O_2 and EC procedures, the usage of the peroxi-EC method demonstrated improved pollutant degrading efficiency. Furthermore, the peroxi-EC process's net energy usage was greatly decreased.

7.3.3 Electrocoagulation—electrooxidation (EC-EO)

Utilizing the oxidation phenomena, electrooxidation (EO) is a water treatment method that successfully eliminates organic compounds from various wastewaters. The EO technique focuses on using hydroxyl radicals to combat contaminants found in wastewaters [42, 43]. When contaminants come into contact with the electrode's surface and become adsorbed there, the oxidation reaction occurs in a direct electrochemical process. The primary advantages of the EO process are the total elimination and disposal of biological sludge, lower maintenance costs, ease of use, and straightforward equipment design. However, the process's application is limited, especially when colloidal particles are present, by prolonged treatment times and comparatively high operational costs [44]. The benefits of utilizing the EC procedure before electrooxidation for the remediation of industrial effluents were examined by Linares-Hernández et al. in 2010 [45]. Two Fe electrodes were attached to a DC power supply as part of the EC procedure during the pretreatment phase. An ideal current density of 80 mA/cm^2 and a treatment duration of 30 minutes were achieved, resulting in the elimination of 87.2% COD, 45.9% BOD, 51.7% color, and 98% total coliform count. In contrast, a Fe electrode was used as the cathode and a boron-doped diamond (BDD) electrode as the anode in the EO method. The treatment duration for the elimination of contaminants was 90 minutes. After 90 minutes of electrolysis, it was found that the COD concentration was completely eliminated, saving about 15% of the total time needed for the complete COD elimination using a stand-alone EO procedure. Additionally, the effluent was completely decolored and all the other pollutants were removed with >99% efficiency using the hybrid EC-EO process. Yılmaz Nayır and Kara (2018) assessed the treatment of wastewater from container washing (CWW) using a hybrid EC-EO process in a different study [46]. The elimination of color and COD content was the primary emphasis of the investigation. Both Al and Fe electrodes were used in the EC process. The Fe electrode produced the greatest color and COD elimination effectiveness of 98% and 82%, respectively, under ideal conditions. The EC method performed less well in removing COD content, though. As a result, the EO procedure was put into place as an EC posttreatment measure. While the decolorization efficiency dropped to 71% at the same wavelength and current density, the COD elimination rate greatly increased to about 90% with the utilization of a BDD electrode during the EO technique. Even though the EO technique needed a comparatively longer time to achieve enhanced pollutant elimination rate, this constraint can be successfully overcome using the hybrid EC-EO process without impairing process performance.

7.3.4 Electrocoagulation—Ultrasound (EC-US)

It is anticipated that the energy produced by ultrasonic (US) irradiation will enhance the EC process's kinetics and pollutant removal efficiency. Additionally, US irradiation greatly improved iron hydroxide removal efficiency [47, 48]. By creating surface imperfections, US radiation stimulates the electrodes as well as the ions that are present in the electrodes' reaction zone. Additionally, US radiation causes local amplification by raising the friction between the liquid and the electrode surface. The energy generated by the US radiation significantly enhances the generation of free radicals, specifically $HO_2^{\bullet}$, $H^{\bullet}$, and $OH^{\bullet}$. These radicals are responsible for the degradation of organic pollutants. As a result, the EC procedure combined with US irradiation has become a viable combined method for the elimination of various harmful contaminants [49]. Chu et al. (2012) investigated how the hybrid US-EC process affected the ability of the wastewater from car washes to reduce turbidity and COD [50]. The hybrid process's removal efficiency was calculated as a function of time. The majority of the pollutant degradation takes place during the first 20 minutes of operation. To assess their impact on the process performance, variables such as reaction time, inter-electrode spacing, and current density were changed. The combined US-EC technique demonstrated COD and turbidity elimination rates of 81% and 95.5%, respectively, under ideal conditions. According to the study's findings, the final lower values of turbidity and COD satisfied the requirements for surface water reuse. Özyonar et al. (2020) used US, EC, and US-EC procedures to remove color and COD from aqueous solution spiked with dispersion blue and reactive red dyes [51]. The US approach was able to remove only 36% of the COD and 50% of the color, whereas the EC process was able to remove 90% of the color and 80% of the COD. With full decolorization, the coupled US-EC technique demonstrated a 99% COD elimination rate. These results confirmed that the hybrid US-EC process performed better than the stand-alone EC and US processes in terms of color and COD removal efficiency.

7.4 INTEGRATED ELECTROCOAGULATION—ADSORPTION PROCESSES

7.4.1 Electrocoagulation—Activated Carbon (EC-AC)

Owing to its extensive surface area and well-developed porous structure, activated carbon (AC) finds widespread application as an adsorbent. To improve the removal of organic contaminants from wastewater, EC and AC can be used in combination [52, 53]. Figure 7.2 displays the hybrid electrocoagulation—activated carbon process. Total organic carbon (TOC) in peat solution might be removed by combining electrocoagulation with activated carbon adsorption according to research by Myllymaki et al. (2018) [54]. Several process configurations, such as EC-AC and AC-EC, were

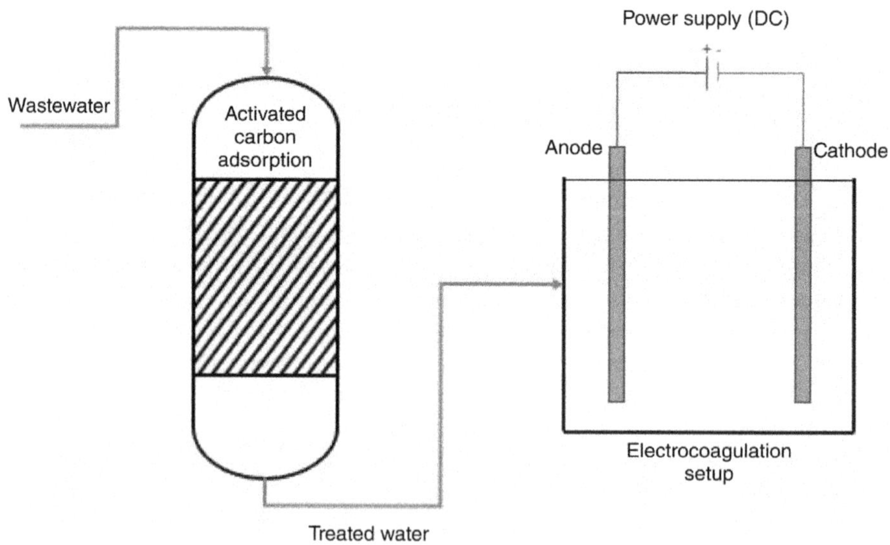

FIGURE 7.2 A schematic representation of the combined EC-AC method (Reproduced with permission from Almukdad et al. (2021) @ Elsevier [2]).

examined. Iron or aluminum electrodes were used as the anode while stainless steel electrodes were used as the cathode in the EC system. They were placed 0.5 cm apart and linked to a DC power supply for a duration of 60 minutes at a current density of 8 mA/cm^2. The TOC removal rates when the iron and aluminum electrodes were used as the anode were 86.7% and 79.7%, respectively. The activated carbon is made up of 92% carbon and 82% mesopores in the adsorption system. With a 10 g/L activated carbon concentration, the maximal TOC elimination achieved using the activated carbon procedure was 90%. With aluminum electrodes in the coupled EC-AC method, the TOC elimination rate reached 94.4%. On the other hand, the TOC elimination rate dropped to 93.3% with the aluminum electrode and 73.3% with the iron electrode in the coupled EC-AC method. Overall, it was discovered that the AC-EC hybrid system was effective in removing organic materials from wastewater.

7.4.2 Electrocoagulation—Biofiltration (EC-BF)

Because biofiltration is easy to use, inexpensive to treat, and very effective at removing nitrogen and biodegradable organic pollutants, it is commonly employed in the treatment of landfill leachate [55, 56]. Figure 7.3 displays the hybrid electrocoagulation—biofiltration system. The effects of employing the EC method as a pretreatment for

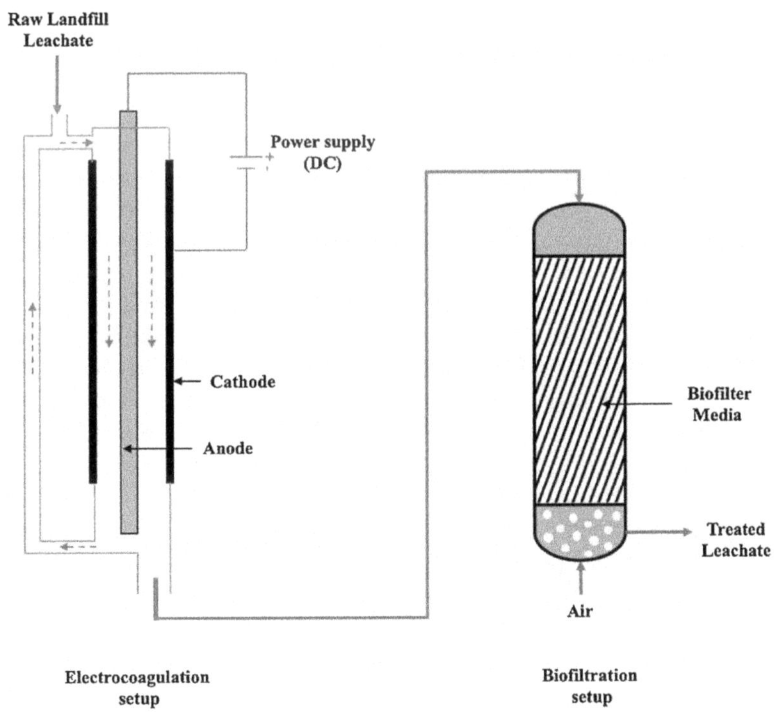

FIGURE 7.3 A schematic representation of the combined EC-BF method (Reproduced with permission from Almukdad et al. (2021) @ Elsevier [2]).

biofiltration (BF) of landfill leachates were investigated by Dia et al. (2018) [57]. The removal rates of COD, phosphorus, NH_4, and metals were examined in order to gauge how effective the combination of these two distinct treatments was. The EC system employed a completely cylindrical rod made of aluminum as the anode and a hollow cylindrical rod made of stainless steel as the cathode. The rods were spaced 1.55 cm apart and connected to a DC power supply. With an initial pH value of 7.83, a current density of 8 mA/cm^2, and a treatment duration of 20 minutes, the maximum COD removal of 82% was achieved. An aerated column filled with a combination of wood and peat chips was utilized to facilitate the biofiltration process. Aeration is a functional sludge processor that encourages microbial growth in wastewater by forcing air into the tank. Following their feeding on the organic waste, the microorganisms cause flocs to develop and sink to the bottom of the container. By applying EC technology as a pretreatment to landfill leachate, the removal rates of color, phosphorus, iron, zinc, and COD increased by 60%, 15%, 82%, and 37%, respectively. With a mere 6% removal rate, the EC technique was shown to be ineffective in removing ammonia. 63.2% of COD, >97% of BOD_5, and 49% of TOC_{Hyl} were removed utilizing the hybrid system. Metals and inorganic contaminants were removed at a rate of above 99%. Specifically, the removal rates of turbidity, color, iron, zinc, phosphorus, and aluminum were 39%, 69.9%, 78.9%, 47.5%, and 96.5%, respectively. The approximate total cost of operating the hybrid process was 1.23 US/m^3 [58].

7.5 SCALING UP OF HYBRID ELECTROCOAGULATION PROCESSES

To apply the technique at an industrial scale, it is crucial to develop a scalable hybrid EC bench scale procedure. A few experiments were carried out at the pilot scale, despite the fact that hybrid electrocoagulation-based methods have demonstrated promising pollutant removal efficiency and low energy usage [59, 60]. The efficacy of EC hybrid systems for treating wastewater from the food processing industry, pretreatment of seawater, and municipal wastewater was assessed in four different experiments. Timmes et al. (2010) used a hybrid EC-UF technique to remediate seawater [61]. The EC-UF procedure has a total volume of 56.68 L. When saltwater was pretreated with EC, fouling in the UF stage was decreased. Because of the increased scouring of hydrogen at the cathode in EC, iron electrodes were utilized, and their polarity was switched after 250 s. This led to a continuous iron dose. The EC procedure yielded an average iron dosage of 0.73 mg/L. The same dose of $FeCl_3$ caused a 20% drop in transmembrane pressure in UF. It was discovered by X-ray diffraction that the electrodes had deposits of magnetite, maghemite, lepidocrocite, and akageneite. The EC unit's ineffective hydraulic design was blamed for the deposition of precipitates on the electrodes. In order to maximize scouring velocity and prevent precipitates from depositing on the electrodes, it was advised that EC units be constructed with a short retention period and a higher current density. Additionally, it was advised to use extended cycling times for polarity reversal. The first electro- bioreactor-based pilot plant for treating municipal wastewater was created by Ibeid et al. (2013) [62]. The reactor's overall capacity was 250 L, and filtration was accomplished using a hollow fiber PVDF membrane module. Two cylindrical electrodes connected to a DC power supply were utilized for EC. 15 A/cm^2 of current density was maintained, and a 5-minute pulse on-time and a 20-minute pulse off-time were used. The reactor's hydraulic retention duration was 11 hours and the average inflow concentration was 3000 mg/L. Combining EC with MBR was proved to minimize membrane fouling by 300%. Membrane fouling in MBR is mostly caused by organic colloids and soluble microbial products (SMP). Membrane fouling was decreased by employing EC to eliminate SMP and organic colloids. Additionally, the suspended solids' morphology and structure were altered by the DC electric field, which improved the electroosmotic extraction of tightly bound water from suspended solids.

7.6 CONCLUSION

The use of the hybrid electrocoagulation technique for water treatment applications has attracted more interest in recent years. It was discovered that employing a hybrid EC approach could improve pollution removal while lowering the overall process costs. The effectiveness of combining the EC process with other water treatment techniques, such as membranes, chemicals, electromagnetic fields, and oxidation processes, is assessed in this chapter. The efficacy of the EC process's pollutant removal is further increased when combined with other water treatment methods. Additionally, it was noted that the hybrid EC procedure significantly lowers the treatment's overall energy consumption and operating costs. The effectiveness of several water treatment techniques, including membrane filtration, oxidation, and adsorption in conjunction with the EC process, is examined in this paper. Because of its minimal environmental impact, simple handling during operation, and ease of use, the hybrid technology that combines membrane filtration with EC process has shown to be one of the most promising techniques for treating wastewater and water. Even though EC combined with other methods achieved excellent process performance, its industrial applicability is still limited by issues such as electrode passivation and non-uniform coagulant dosage. Alternate current can be used to remedy this, though, as can routine electrode cleaning. To lower electrode consumption and improve the process's efficiency in reducing pollutants, more research needs to be carried out. This might be accomplished by assessing the effectiveness of dielectrophilic enhanced electrodes, which are expected to reduce electrode usage and enhance the permeate water quality. Such information would help in developing future reactor design of the combined approaches as well as in understanding the mechanism of pollutant removal and fouling control by the EC process.

REFERENCES

[1] P.P. Das, N.S. Samanta, S. Dhara, M.K. Purkait, Biofuel production from algal biomass, in: M. Shah (Ed.), *Green Approach to Alternative Fuel for a Sustainable Future*, Elsevier, 2023: pp. 167–179. https://doi.org/10.1016/B978-0-12-824318-3.00009-6.

[2] A. Almukdad, M.A. Hafiz, A.T. Yasir, R. Alfahel, A.H. Hawari, Unlocking the application potential of electrocoagulation process through hybrid processes, *J. Water Process Eng.* 40 (2021) 101956. https://doi.org/10.1016/j.jwpe.2021.101956.

[3] P.P. Das, M. Sharma, M.K. Purkait, Recent progress on electrocoagulation process for wastewater treatment: A review, *Sep. Purif. Technol.* 292 (2022) 121058. https://doi.org/10.1016/j.seppur.2022.121058.

[4] M. Sharma, P.P. Das, T. Sood, A. Chakraborty, M.K. Purkait, Reduced graphene oxide incorporated polyvinylidene fluoride/cellulose acetate proton exchange membrane for energy extraction using microbial fuel cells, *J. Electroanal. Chem.* 907 (2022) 115890. https://doi.org/10.1016/j.jelechem.2021.115890.
[5] M. Sharma, P.P. Das, T. Sood, A. Chakraborty, M.K. Purkait, Ameliorated polyvinylidene fluoride based proton exchange membrane impregnated with graphene oxide, and cellulose acetate obtained from sugarcane bagasse for application in microbial fuel cell, *J. Environ. Chem. Eng.* 9 (2021) 106681. https://doi.org/10.1016/j.jece.2021.106681.
[6] N.S. Samanta, P.P. Das, P. Mondal, U. Bora, M.K. Purkait, Physico-chemical and adsorption study of hydrothermally treated zeolite A and FAU-type zeolite X prepared from LD (Linz—Donawitz) slag of the steel industry, *Int. J. Environ. Anal. Chem.* (n.d.) 1–23. https://doi.org/10.1080/03067319.2022.2079082.
[7] N.S. Samanta, P.P. Das, P. Mondal, M. Changmai, M.K. Purkait, Critical review on the synthesis and advancement of industrial and biomass waste-based zeolites and their applications in gas adsorption and biomedical studies, *J. Indian Chem. Soc.* 99 (2022) 100761. https://doi.org/10.1016/j.jics.2022.100761.
[8] M. Changmai, P.P. Das, P. Mondal, M. Pasawan, A. Sinha, P. Biswas, S. Sarkar, M.K. Purkait, Hybrid electrocoagulation—microfiltration technique for treatment of nanofiltration rejected steel industry effluent, *Int. J. Environ. Anal. Chem.* 102 (2022) 62–83. https://doi.org/10.1080/03067319.2020.1715381.
[9] P.P. Das, A.D. Sontakke, M.K. Purkait, Rice straw for biofuel production, in: M.P. Shah (Ed.), *Green Approach to Alternative Fuel for a Sustainable Future*, Elsevier, 2023: pp. 153–166. https://doi.org/10.1016/B978-0-12-824318-3.00034-5.
[10] P.P. Das, P. Mondal, Anweshan, A. Sinha, P. Biswas, S. Sarkar, M.K. Purkait, Treatment of steel plant generated biological oxidation treated (BOT) wastewater by hybrid process, *Sep. Purif. Technol.* 258 (2021). https://doi.org/10.1016/j.seppur.2020.118013.
[11] P.P. Das, A. Anweshan, P. Mondal, A. Sinha, P. Biswas, S. Sarkar, M.K. Purkait, Integrated ozonation assisted electrocoagulation process for the removal of cyanide from steel industry wastewater, *Chemosphere* 263 (2021) 128370. https://doi.org/10.1016/j.chemosphere.2020.128370.
[12] P.P. Das, P. Mondal, Membrane-assisted potable water reuses applications: Benefits and drawbacks, in: M. Sillanpää, A. Khadir, K. Gurung (Eds.), *Resource Recovery in Drinking Water Treatment*, Elsevier, 2023: pp. 289–309. https://doi.org/10.1016/B978-0-323-99344-9.00014-1.
[13] P.P. Das, A. Anweshan, M.K. Purkait, Treatment of cold rolling mill (CRM) effluent of steel industry, *Sep. Purif. Technol.* 274 (2021) 119083. https://doi.org/10.1016/j.seppur.2021.119083.
[14] P.P. Das, A.D. Sontakke, M.K. Purkait, Electrocoagulation process for wastewater treatment: Applications, challenges, and prospects, in: M.P. Shah (Ed.), *Development in Wastewater Treatment Research and Processes*, Elsevier, 2023: pp. 23–48. https://doi.org/10.1016/B978-0-323-95684-0.00015-4.
[15] P.P. Das, S. Dhara, M.K. Purkait, Hybrid electrocoagulation and ozonation techniques for industrial wastewater treatment, in: M.P. Shah (Ed.), *Sustainable Industrial Wastewater Treatment and Pollution Control*, Springer Nature Singapore, Singapore, 2023: pp. 107–128. https://doi.org/10.1007/978-981-99-2560-5_6.
[16] M. Sharma, P.P. Das, M.K. Purkait, Energy storage properties of nanomaterials, in: A. Husen, K.S. Siddiqi (Eds.), *Advances in Smart Nanomaterials and their Applications*, Elsevier, 2023: pp. 337–350. https://doi.org/10.1016/B978-0-323-99546-7.00005-7.
[17] M. Sharma, P.P. Das, A. Chakraborty, M.K. Purkait, Clean energy from salinity gradients using pressure retarded osmosis and reverse electrodialysis: A review, *Sustain. Energy Technol. Assessments* 49 (2022) 101687. https://doi.org/10.1016/j.seta.2021.101687.

[18] A. Bagga, S. Chellam, D.A. Clifford, Evaluation of iron chemical coagulation and electrocoagulation pretreatment for surface water microfiltration, *J. Memb. Sci.* 309 (2008) 82–93. https://doi.org/10.1016/j.memsci.2007.10.009.

[19] N.P. Gamage, J.D. Rimer, S. Chellam, Improvements in permeate flux by aluminum electroflotation pretreatment during microfiltration of surface water, *J. Memb. Sci.* 411–412 (2012) 45–53. https://doi.org/10.1016/j.memsci.2012.04.014.

[20] M. Changmai, M. Pasawan, M.K. Purkait, Treatment of oily wastewater from drilling site using electrocoagulation followed by microfiltration, *Sep. Purif. Technol.* 210 (2019) 463–472. https://doi.org/10.1016/j.seppur.2018.08.007.

[21] W.B.J.S.P. Konieczny, Protein and water recovery from poultry processing wastewater integrating microfiltration, ultrafiltration and vacuum membrane distillation, *Int. J. Environ. Sci. Technol.* 12 (2015) 1875–1888. https://doi.org/10.1007/s13762-014-0557-4.

[22] N. Shekhar Samanta, P.P. Das, S. Dhara, M.K. Purkait, An overview of precious metal recovery from steel industry slag: Recovery strategy and utilization, *Ind. Eng. Chem. Res.* 62 (2023) 9006–9031. https://doi.org/10.1021/acs.iecr.3c00604.

[23] M. Sharma, P.P. Das, A. Chakraborty, M.K. Purkait, Extraction of clean energy from industrial wastewater using bioelectrochemical process, in: M. Sillanpää, A. Khadir, K. Gurung (Eds.), *Resource Recovery in Industrial Waste Waters*, Elsevier, 2023: pp. 601–620. https://doi.org/10.1016/B978-0-323-95327-6.00003-8.

[24] K. Sardari, J. Askegaard, Y.H. Chiao, S. Darvishmanesh, M. Kamaz, S.R. Wickramasinghe, Electrocoagulation followed by ultrafiltration for treating poultry processing wastewater, *J. Environ. Chem. Eng.* 6 (2018) 4937–4944. https://doi.org/10.1016/j.jece.2018.07.022.

[25] O. Yahiaoui, H. Lounici, N. Abdi, N. Drouiche, N. Ghaffour, A. Pauss, N. Mameri, Treatment of olive mill wastewater by the combination of ultrafiltration and bipolar electrochemical reactor processes, *Chem. Eng. Process. Process Intensif.* 50 (2011) 37–41. https://doi.org/10.1016/j.cep.2010.11.003.

[26] A.D. Sontakke, P.P. Das, P. Mondal, M.K. Purkait, Thin-film composite nanofiltration hollow fiber membranes toward textile industry effluent treatment and environmental remediation applications: Review, *Emergent Mater.* (2021). https://doi.org/10.1007/s42247-021-00261-y.

[27] T. Tavangar, K. Jalali, M.A. Alaei Shahmirzadi, M. Karimi, Toward real textile wastewater treatment: Membrane fouling control and effective fractionation of dyes/inorganic salts using a hybrid electrocoagulation—nanofiltration process, *Sep. Purif. Technol.* 216 (2019) 115–125. https://doi.org/10.1016/j.seppur.2019.01.070.

[28] Z.B. Gönder, G. Balcıoğlu, I. Vergili, Y. Kaya, An integrated electrocoagulation—nanofiltration process for carwash wastewater reuse, *Chemosphere* 253 (2020). https://doi.org/10.1016/j.chemosphere.2020.126713.

[29] P.P. Das, Deepti, M.K. Purkait, Industrial wastewater to biohydrogen production via potential bio-refinery route, in: M.P. Shah (Ed.), *Biorefinery for Water and Wastewater Treatment*, Springer International Publishing, Cham, 2023: pp. 159–179. https://doi.org/10.1007/978-3-031-20822-5_8.

[30] M. Bharti, P.P. Das, M.K. Purkait, A review on the treatment of water and wastewater by electrocoagulation process: Advances and emerging applications, *J. Environ. Chem. Eng.* 11 (2023) 111558. https://doi.org/10.1016/j.jece.2023.111558.

[31] S. Zhao, G. Huang, G. Cheng, Y. Wang, H. Fu, Hardness, COD and turbidity removals from produced water by electrocoagulation pretreatment prior to reverse osmosis membranes, *Desalination* 344 (2014) 454–462. https://doi.org/10.1016/j.desal.2014.04.014.

[32] K. Sadeddin, A. Naser, A. Firas, Removal of turbidity and suspended solids by electro-coagulation to improve feed water quality of reverse osmosis plant, *Desalination* 268 (2011) 204–207. https://doi.org/10.1016/j.desal.2010.10.027.

[33] J.N. Hakizimana, B. Gourich, C. Vial, P. Drogui, A. Oumani, J. Naja, L. Hilali, Assessment of hardness, microorganism and organic matter removal from seawater by electrocoagulation as a pretreatment of desalination by reverse osmosis, *Desalination* 393 (2016) 90–101. https://doi.org/10.1016/j.desal.2015.12.025.

[34] P.P. Das, S. Dhara, M.K. Purkait, Ozone-based oxidation processes for the removal of pharmaceutical products from wastewater, in: M.P. Shah, P. Ghosh (Eds.), *Development in Wastewater Treatment Research and Processes*, Elsevier, 2024: pp. 287–308. https://doi.org/10.1016/B978-0-443-19207-4.00003-3.

[35] M. Sharma, P.P. Das, S. Kumar, M.K. Purkait, Polyurethane foams as packing and insulating materials, in: *Polyurethanes: Preparation, Properties, and Applications*, Vol. *Emerging Applications*, American Chemical Society, 2023: pp. 4–83. https://doi.org/10.1021/bk-2023-1454.ch004.

[36] L. Bilińska, K. Blus, M. Gmurek, S. Ledakowicz, Coupling of electrocoagulation and ozone treatment for textile wastewater reuse, *Chem. Eng. J.* 358 (2019) 992–1001. https://doi.org/10.1016/j.cej.2018.10.093.

[37] M.A. Ahangarnokolaei, B. Ayati, H. Ganjidoust, Simultaneous and sequential combination of electrocoagulation and ozonation by Al and Fe electrodes for DirectBlue71 treatment in a new reactor: Synergistic effect and kinetics study, *Chemosphere* 285 (2021) 131424. https://doi.org/10.1016/j.chemosphere.2021.131424.

[38] Z. Qiang, J.H. Chang, C.P. Huang, Electrochemical regeneration of Fe2+ in Fenton oxidation processes, *Water Res.* 37 (2003) 1308–1319. https://doi.org/10.1016/S0043-1354(02)00461-X.

[39] S. Dhara, N. Shekhar Samanta, P.P. Das, R.V.S. Uppaluri, M.K. Purkait, Ravenna grass-extracted alkaline lignin-based polysulfone mixed matrix membrane (MMM) for Aqueous Cr(VI) removal, *ACS Appl. Polym. Mater.* 5 (2023) 6399–6411. https://doi.org/10.1021/acsapm.3c00999.

[40] P. Asaithambi, B. Sajjadi, A.R. Abdul Aziz, W.M.A. Bin Wan Daud, Performance evaluation of hybrid electrocoagulation process parameters for the treatment of distillery industrial effluent, *Process Saf. Environ. Prot.* 104 (2016) 406–412. https://doi.org/10.1016/j.psep.2016.09.023.

[41] E. Yüksel, I.A. Şengil, M. Özacar, The removal of sodium dodecyl sulfate in synthetic wastewater by peroxi-electrocoagulation method, *Chem. Eng. J.* 152 (2009) 347–353. https://doi.org/10.1016/j.cej.2009.04.058.

[42] E. Brillas, C.A. Martínez-Huitle, Decontamination of wastewaters containing synthetic organic dyes by electrochemical methods. An updated review, *Appl. Catal. B Environ.* 166–167 (2015) 603–643. https://doi.org/10.1016/j.apcatb.2014.11.016.

[43] M. Panizza, G. Cerisola, Direct and mediated anodic oxidation of organic pollutants, *Chem. Rev.* 109 (2009) 6541–6569. https://doi.org/10.1021/cr9001319.

[44] S. Chakraborty, P.P. Das, P. Mondal, Recent advances in membrane technology for the recovery and reuse of valuable resources, in: M. Sillanpää, A. Khadir, K. Gurung (Eds.), *Resource Recovery in Industrial Waste Waters*, Elsevier, 2023: pp. 695–719. https://doi.org/10.1016/B978-0-323-95327-6.00028-2.

[45] I. Linares-Hernández, C. Barrera-Díaz, B. Bilyeu, P. Juárez-GarcíaRojas, E. Campos-Medina, A combined electrocoagulation-electrooxidation treatment for industrial wastewater, *J. Hazard. Mater.* 175 (2010) 688–694. https://doi.org/10.1016/j.jhazmat.2009.10.064.

[46] T. Yılmaz Nayır, S. Kara, Container washing wastewater treatment by combined electrocoagulation—electrooxidation, *Sep. Sci. Technol.* 53 (2018) 1592–1603. https://doi.org/10.1080/01496395.2017.1411365.

[47] Z. Al-qodah, M. Al-shannag, Z. Al-qodah, M. Al-shannag, On the performance of free radicals combined electrocoagulation treatment processes, *Separat. Purif. Rev.* (2019). https://doi.org/10.1080/15422119.2018.1459700.

[48] P.P. Das, P. Mondal, M.K. Purkait, Recent advances in synthesis of iron nanoparticles via green route and their application in biofuel production, in: M. Srivastava, M.A. Malik, P.K. Mishra (Eds.), *Green Nano Solution for Bioenergy Production Enhancement*, Springer Nature Singapore, Singapore, 2022: pp. 79–104. https://doi.org/10.1007/978-981-16-9356-4_4.
[49] P.P. Das, A.D. Sontakke, N.S. Samanta, M.K. Purkait, Emerging contaminants in wastewater: Eco-toxicity and sustainability assessment, in: M.P. Shah (Ed.), *Industrial Wastewater Reuse: Applications, Prospects and Challenges*, Springer Nature Singapore, Singapore, 2023: pp. 63–87. https://doi.org/10.1007/978-981-99-2489-9_4.
[50] J.Y. Chu, Y.R. Li, N. Li, W.H. Huang, Treatment of car-washing wastewater by electrocoagulation-ultrasound technique for reuse, *Adv. Mater. Res.* 433–440 (2012) 227–232. https://doi.org/10.4028/www.scientific.net/AMR.433-440.227.
[51] F. Özyonar, Ö. Gökkuş, M. Sabuni, Removal of disperse and reactive dyes from aqueous solutions using ultrasound-assisted electrocoagulation, *Chemosphere* 258 (2020). https://doi.org/10.1016/j.chemosphere.2020.127325.
[52] A. Bhatnagar, W. Hogland, M. Marques, M. Sillanpää, An overview of the modification methods of activated carbon for its water treatment applications, *Chem. Eng. J.* 219 (2013) 499–511. https://doi.org/10.1016/j.cej.2012.12.038.
[53] N.S. Samanta, P. P. Das, M. Sharma, M.K. Purkait, Recycle of water treatment plant sludge and its utilization for wastewater treatment, in: M. Sillanpää, A. Khadir, K. Gurung (Eds.), *Resource Recovery in Drinking Water Treatment*, Elsevier, 2023: pp. 239–264. https://doi.org/10.1016/B978-0-323-99344-9.00010-4.
[54] P. Myllymäki, R. Lahti, H. Romar, U. Lassi, Removal of total organic carbon from peat solution by hybrid method—electrocoagulation combined with adsorption, *J. Water Process Eng.* 24 (2018) 56–62. https://doi.org/10.1016/j.jwpe.2018.05.008.
[55] T.A. Kurniawan, W.H. Lo, G.Y.S. Chan, Physico-chemical treatments for removal of recalcitrant contaminants from landfill leachate, *J. Hazard. Mater.* 129 (2006) 80–100. https://doi.org/10.1016/j.jhazmat.2005.08.010.
[56] P.P. Das, S. Dhara, M.K. Purkait, The anaerobic ammonium oxidation process: Inhibition, challenges and opportunities, in: M.P. Shah (Ed.), *Ammonia Oxidizing BacteriaApplications in Industrial Wastewater Treatment*, Royal Society of Chemistry, 2023: pp. 56–82. https://doi.org/10.1039/BK9781837671960-00056.
[57] O. Dia, P. Drogui, G. Buelna, R. Dubé, Hybrid process, electrocoagulation-biofiltration for landfill leachate treatment, *Waste Manag.* 75 (2018) 391–399. https://doi.org/10.1016/j.wasman.2018.02.016.
[58] D. Huang, L. Yang, W. Xu, Q. Chen, J.H. Ko, Q. Xu, Enhancement of the methane removal efficiency via aeration for biochar-amended landfill soil cover, *Environ. Pollut.* 263 (2020) 114413. https://doi.org/10.1016/j.envpol.2020.114413.
[59] S. Khoufi, F. Feki, S. Sayadi, Detoxification of olive mill wastewater by electrocoagulation and sedimentation processes, *J. Hazard. Mater.* 142 (2007) 58–67. https://doi.org/10.1016/j.jhazmat.2006.07.053.
[60] P.P. Das, M. Sharma, M.K. Purkait, Ammonia oxidizing bacteria in wastewater treatment, in: M.P. Shah (Ed.), *Ammonia Oxidizing Bacteria: Applications in Industrial Wastewater Treatment*, Royal Society of Chemistry, 2023: pp. 83–102. https://doi.org/10.1039/BK9781837671960-00083.
[61] T.C. Timmes, H.-C. Kim, B.A. Dempsey, Electrocoagulation pretreatment of seawater prior to ultrafiltration: Pilot-scale applications for military water purification systems, *Desalination* 250 (2010) 6–13. https://doi.org/10.1016/j.desal.2009.03.021.
[62] S. Ibeid, M. Elektorowicz, J.A. Oleszkiewicz, Novel electrokinetic approach reduces membrane fouling, *Water Res.* 47 (2013) 6358–6366. https://doi.org/10.1016/j.watres.2013.08.007.

8 Case Studies on Electrocoagulation Treatment of Water and Wastewater

8.1 INTRODUCTION

Numerous industrial operations utilize significant quantities of fresh water, resulting in the generation of wastewater. It is essential to appropriately treat this wastewater to reduce or eliminate pollutants and achieve the necessary purity levels for reuse in industrial processes. However, there is a demand for innovative, cost-effective methods to purify water for human consumption and to treat wastewater from industrial effluents before releasing it into other water systems. The treatment approach varies based on the specific characteristics of wastewater and the pollutants present. Traditional methods in wastewater treatment encompass advanced oxidation processes [1], biological processes [2], physicochemical processes as well as emerging technologies such as membrane filtration and adsorption. Advanced oxidation processes necessitate potent oxidants, which can increase safety concerns and costs in wastewater treatment. Conversely, biological processes require tightly controlled conditions, long retention times, and larger physical space, and they may lead to the generation of undesirable by-products. Chemical processes involve significant chemical inputs, increasing process costs and complexity in downstream operations, thereby increasing the risk of secondary contamination. While membrane filtration [3] and adsorption have their merits, they typically require comprehensive pretreatment to efficiently treat wastewater; otherwise, they may lose effectiveness over time owing to pore blockage by contaminants, resulting in reduced flow rates of clean treated effluent per unit surface area. Consequently, electrochemical processes have attracted substantial interest in wastewater treatment research. Among these processes, EC is a simple and remarkably efficient method for treating various wastewater types from a variety of industry effluents, including food, pharmaceutical, textile, tannery, and pulp and paper industries [4]. This process involves applying direct current (DC) via two sacrificial electrodes that are submerged in the treating solution. This dissolves the sacrificial anode (electrical oxidation), which is formed by soluble and insoluble species (hydroxide metal). The coagulant acts as a means of adsorbing and removing soluble or colloidal substances through sedimentation or flotation. Coagulant ions are continuously produced and generated *in situ* during the EC procedure. The efficacy of the EC method is influenced by several variables, including pH, current density (CD), electrode materials, electrolytic duration, and electrode spacing. EC is a quick

DOI: 10.1201/9781003499671-8

and easy way to purify wastewater from several industrial effluents and drinking water. The anodized aluminum (produced by the electrical dissolution of the anode) is more efficient than the aluminum ion incorporated in the form of aluminum sulfate solution [5]. Numerous researchers globally have conducted numerous EC studies to date, targeting different types of wastewater, yielding promising results. Nevertheless, ongoing advancements in EC research have significantly emphasized the importance of EC and its potential positive effects on wastewater treatment. EC process emerges as the most environmentally sustainable option for treating wastewater due to its uncomplicated installation, minimal space requirements, and capacity to handle large volumes of water without extensive chemical intervention [6]. However, this chapter examines the latest developments in the evolution of EC and its emerging hybrid processes besides providing a comprehensive overview of several case studies on the treatment of water and wastewater.

8.2 ADVANCES IN BATCH AND CONTINUOUS ELECTROCOAGULATION REACTORS

Advances in EC reactors have significantly improved the treatment of various wastewaters, particularly in batch and continuous operations. In batch EC, recent advancements focus on enhancing efficiency and scalability. In continuous EC, reactors have also seen notable advancements. Automation and control systems have been integrated to optimize process parameters in real time, ensuring consistent treatment performance and reducing operator intervention. Numerous studies have focused on enhancing the design of electrochemical reactors to improve their performance in treating wastewater. Figure 8.1 illustrates the various reactor designs employed in the EC method [4]. Fouad et al. (2009) explored the efficacy of horizontal and vertical electrode designs in separating oil and grease during the EC process [7]. They used a configuration of horizontally arranged aluminum cylinders as the anode cell and found that this horizontal design outperformed the traditional vertical parallel plate cell. Another study investigated the use of cylindrical rotating electrodes to demulsify the emulsion of crude oil in the EC process. This approach reduced anode passivation by enhancing mass transfer rates between the anode and the solution. The turbulence generated by the cylindrical rotating electrode also improved mixing in the bulk solution, enhancing process effectiveness. Furthermore, achieving the separation of oil and grease in electrochemical cells can be accomplished by developing high-porosity fixed-bed sacrificial anodes. These advancements in reactor design significantly contribute to optimization of EC processes for wastewater treatment. It offers several advantages compared to modern electrode designs (horizontal or vertical anodes), containing a smaller spatial footprint, a higher removal rate, and extended treatment intervals between electrode replacements. Researchers explored the utilization of an Al wire gauze anode in a batch electrochemical cell for separating crude oil emulsions [8]. A different reactor design for treating oil-in-water emulsions involves employing dimensionally stable electrodes (anodes), leveraging electrochemical oxidation, a process absent in standard EC methods. Furthermore, improvements in managing

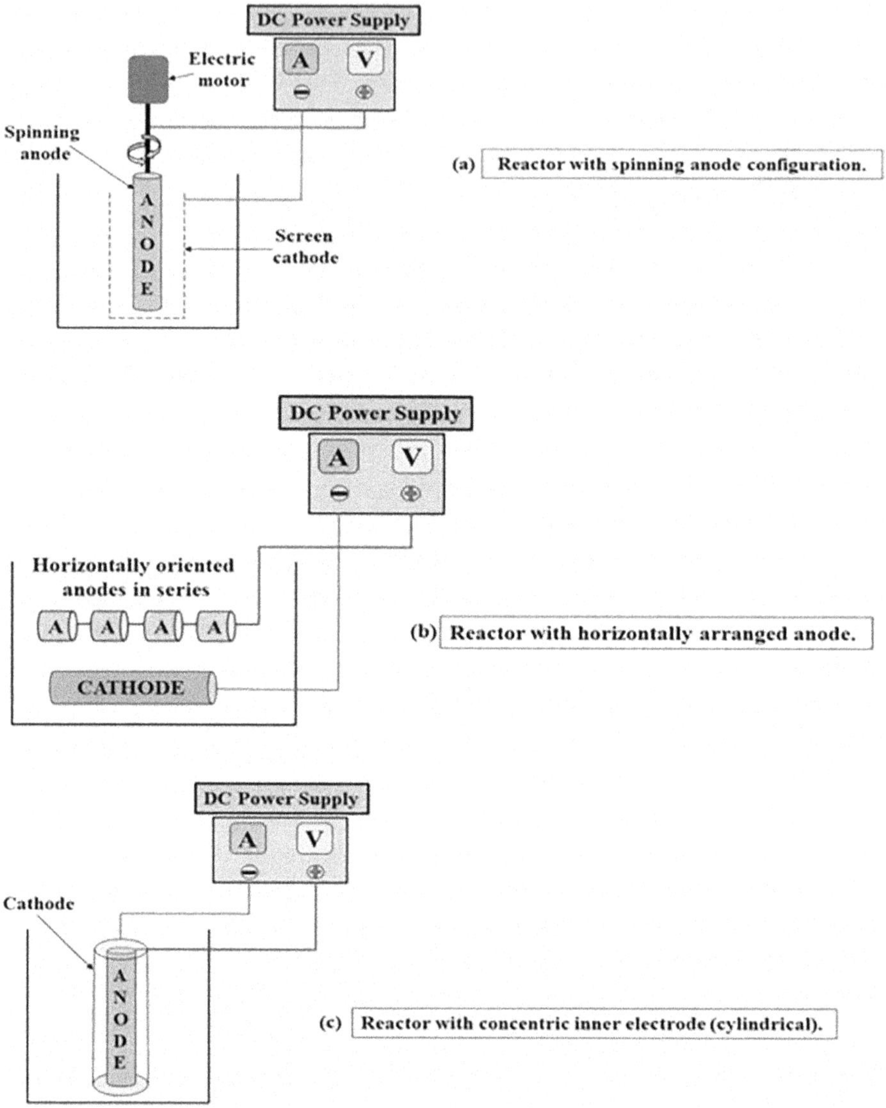

FIGURE 8.1 Types of electrocoagulation reactor designs (Reproduced with permission from Bharti et al. (2023) @ Elsevier [4]).

process parameters have greatly enhanced the dynamics of the EC process in recent times. For instance, Genc and Bakirci (2015) assessed how pulse electric voltage affected energy consumption while separating oil emulsions using the EC method. The application of pulse electric voltage was noted to effectively reduce the energy consumption rates. During continuous operation, the research demonstrated a notable energy-saving efficiency of 48% [9]. Maha Lakshmi and Sivashanmugam

(2013) employed an integrated sono-EC process to treat oil tanning wastewater. They noted that cavitation bubbles generated during ultra-sonication were easily disrupted or merged, hindering floc formation. This observation led to the inference that continuous sonication adversely impacted the EC process. Nevertheless, adjusting operative control using intermittent ultrasonic impingement was found to effectively address the passivation issue [10].

The design of reactors like rotating cylinders, horizontal and vertical plates, and fixed beds with high porosity has attracted significant attention in recent years for wastewater contaminant removal. The effective functioning of these reactors hinges significantly on several factors including electrode material, reactor volume, operational mode (batch or continuous), depth-to-width (D/W) ratio, total effective area, and power supply. Consequently, optimizing these parameters is essential for the efficient performance of the EC technique. During the operation of the EC method, electrode passivation may occur in direct current mode. This can lead to the obstruction of active sites on the electrodes and a decrease in process efficiency. Nonetheless, this challenge can be addressed by employing alternating current (AC) or pulse electric fields in the solution. For optimizing and modeling the EC process, response surface methodology (RSM) is a statistical and mathematical framework employed for experimental problems [11]. It is well-suited for investigating and optimizing relevant parameters in the EC system. For instance, Zhao et al. (2014) optimized the EC method using RSM and noted that under ideal conditions of a solution pH of 7.35, a CD of 5.90 mA/cm^2, and a treatment duration of 31 minutes, contaminant removal efficiencies were 85.8%, 93.8% and 66.5% for hardness, turbidity, and COD, respectively [12]. In contrast, Ulucan et al. (2014) optimized the EC procedure for treating bilge wastewater and attained the highest removal efficiency of 81.7% for oil and grease at optimum CD and inlet temperature [13]. Chavalparit and Ongwandee (2009) used a Box-Behnken design to eliminate oil content and COD from biodiesel wastewater via the EC technique and achieved a reduction of 55.5% for COD and 98.5% for oil content. These studies demonstrate the effectiveness of RSM in optimizing and analyzing EC processes [14].

8.3 REAL-LIFE APPLICATIONS OF THE ELECTROCOAGULATION PROCESS

EC process finds real-life applications across various industries and environmental settings due to its effectiveness in treating water and wastewater. The versatility and efficiency of the EC process make it a valuable technology with widespread applications in improving water quality, environmental sustainability, and public health.

8.3.1 Treatment of Wine Factory Effluent Using Electrocoagulation Technique

Wastewater generated at wineries primarily comes from cleaning equipment, packaging materials, and surfaces during specific winemaking processes. Various studies have provided literature values for certain parameters found in different kinds of winery wastewater. The composition and volume of effluent can vary significantly depending on the production stage and time of year. The largest volume of wastewater occurs during grape harvesting and crushing, followed by fermentation (during the vintage period). These effluents contain a complex mix of organic pollutants such as organic acids (tartaric, lactic, and acetic acids), sugars (glucose and fructose), and alcohols (ethanol and glycerol), as well as persistent high-molecular-weight compounds like polyphenols, tannins, and lignins [15]. The high concentration of organic acids contributes to a low pH value, while the presence of sugars mainly contributes to high COD (chemical oxygen demand) values. Wastewater released during the fermentation of wine contains a high concentration of organic substances. If this water is released into the surroundings without being properly treated, it may significantly disrupt oxygen levels in surface water bodies, which in turn may negatively impact many aquatic creatures. Many treatment techniques have been developed to reduce this risk, but they have differing degrees of success in eliminating other pollutants and organic compounds (COD removal ranges from 25% to 84.5%). Although these methods begin with different approaches, they all involve a combination of various techniques such as photolysis/photo-catalysis, ozone treatment [16], advanced oxidation, constructed wetlands, EC, and biological treatment, followed by coagulation, and typically require a relatively long treatment duration. There are various options for treating winery wastewater, and among these, the EC treatment method is acknowledged as a viable alternative for effectively breaking down wastewater with high organic content. In recent years, the EC method has been employed for treating wastewater from various sectors of the food industry, including distillery and fermentation [17], dairy processing [18], potato chip manufacturing, pasta and cookie production, poultry slaughterhouses, and yeast production [19]. While there is extensive literature detailing various treatment methods, there is limited research focusing on the technical and economic analysis of the EC process, specifically for winery wastewater. Therefore, this study aims at assessing the effectiveness of the EC procedure using Fe and Al electrodes on winery wastewater, with a particular focus on the anodic passivation coating's impact on the overall EC efficiency.

The wastewater analyzed in this study was sourced from an equalization tank at a wine factory situated in Tekirdağ, Turkey. This factory generates an estimated 350 m^3 of wastewater per day. Detailed characteristics of the wastewater are outlined in Table 8.1.

The batch EC reactor made of Plexiglass which had dimensions of 130 mm × 130 mm × 120 mm was used in this investigation. Two electrodes made of Al and Fe with an effective area of 143 cm^2 and a 20 mm separation distance were utilized. Different electrode arrangements were used to link these electrodes to a DC power source. In the anodic passivation coating (APC) investigations, a time relay was used in conjunction with a DC power supply for changing the polarity of the electrodes. The initial solution pH has a major effect on how well the EC process works. For the Al electrode, the monomeric and polymeric $Al(OH)_3$ species have an efficient precipitation mechanism at pH 4.0–6.5, whereas the $Al(OH)_3$ and polymeric $Al(OH)_3$ species have an efficient adsorption mechanism at pH > 6.5 for the elimination of

TABLE 8.1
The properties of the winery wastewater examined in this study (Reproduced from Eyvaz et al. (2014) @ IntechOpen [20]).

Parameter	Value
pH	5.2
COD, mg/L	20400 ± 1100
BOD_5, mg/L	11120 ± 1055
TOC, mg/L	4230 ± 940
TSS, mg/L	1045 ± 85
Turbidity, NTU	1600 ± 510
Color, Pt-Co	5300 ± 100
Conductivity, μS/cm	2800 ± 93

pollutants. Nevertheless, in the case of Fe electrodes, effective removal efficiency can be attained through the formation of flocs within the pH range of 6 to 8. The impact of pH was studied under a constant CD of 40 mA/cm^2 and a treatment period of 60 minutes. No significant differences were observed between the two different electrode materials. The maximum COD elimination efficiencies were seen at a pH of 5 for Al electrodes and at a pH of 7 for Fe electrodes. At these pH levels, maximum turbidity and color removal were achieved, along with minimal energy and electrode consumption and reduced sludge generation. The study noted that COD removal efficiency increased from pH 4 to 5 and then declined at higher pH levels up to 9. Similarly, the Fe electrode exhibited optimal COD elimination performance at pH 7, with maximum turbidity and color removal occurring at pH 5 and 7. The destabilization of colloidal particles was attributed to metal ions generated by anodic dissolution, which then reacted with organic pollutants through adsorption or co-precipitation, precipitating as hydroxides at these specific pH values [21]. In addition, they examined the influence of CD on COD elimination. The elimination rates peaked at a CD range from 40 to 50 mA/cm^2 and either remained constant or decreased at higher CD values. An increase in CD led to the generation of hydroxide flocs, enhancing elimination efficiency through coagulation and resulting in higher energy utilization. However, beyond a certain threshold of CD, cathode passivation occurred, reducing or halting the anode material's dissolution and diminishing floc formation and removal efficiencies. Turbidity and color removal efficacies followed a similar trend to COD elimination efficiency with changes in CD. Based on the experimental findings, they concluded that the APC system can achieve higher removal efficiencies within the same or shorter operation times. Specifically, under identical operating conditions, the APC system yields a 40% greater COD removal than DC. Additionally, APC achieves a 30% faster COD removal performance than DC. Therefore, it can be inferred that anode-cathode polarization decreases the electrolysis time required for metal hydroxides to remove pollutants [20].

TABLE 8.2
Properties of the textile factory wastewater (Alcovel, Béjaia) (Reproduced with permission from Merzouk et al. (2010) @ Elsevier [22]).

Variables	Value
pH	7.86
BOD_5, mg/L	210
COD, mg/L	340
SS, mg/L	300
Turbidity, NTU	130
Color, Pt-Co	>200
Conductivity	3120

8.3.2 Electrocoagulation Treatment of Textile Industry Wastewater

This study utilized the wastewater sourced from a textile industry (Alcovel) located in Béjaia, Algeria, which generates around 5000 m^3 of wastewater per day. The properties of the textile factory effluent are shown in Table 8.2.

This study utilized the wastewater which was sourced from the textile industry located in Béjaia, Algeria, with a daily wastewater output of approximately 5000 m^3. The primary contaminants identified in the effluent were BOD, COD, color, TSS, and turbidity. A 1.5 L capacity EC reactor was constructed for conducting an integrated EC and electrofloatation (EF) process. This reactor housed two rectangular Al electrodes, each with dimensions of 2.7 cm × 1.7 cm × 0.1 cm and a surface area of 4.59 cm^2 and positioned in the middle of the EC reactor horizontally. The removal efficiencies for TSS and turbidity were found to be 85.5% and 76.2%, respectively, under operating conditions of an inter-electrode space ranging from 10 to 30 mm, a treatment time of 10 minutes and a varying CD value ranging from 11.5 to 91.5 mA/cm^2. Furthermore, the concentration reduction was observed at 79.7% for COD and 88.9% for BOD. In addition, a decolonization efficiency of more than 93% was achieved.

8.3.3 Electrocoagulation Process for the Treatment of Cold Rolling Mill (CRM) Wastewater from the Steel Industry

The cold rolling mill (CRM) effluent was collected from the Tata Steel Industry, India. The CRM effluent was characterized, and multiple water quality parameters were assessed, as shown in Table 8.3.

A batch EC system with a capacity of 0.5 L was constructed using an acrylic sheet for conducting the EC process. Aluminum sheets with a suitable surface area of 24.5 cm^2 and dimensions of 7 cm × 3.5 cm were utilized to create both the cathodes and the anodes [24]. In the course of the evaluation of various CDs including 100, 150, and 200 A/m^2, it was noted that experiments accompanied at CDs exceeding 200 A/m^2 exhibited a consistent reduction in contaminant concentration, indicating that the elimination ability of the process reached saturation. Conversely,

TABLE 8.3
The characteristics of cold rolling mill (CRM) wastewater (Reproduced with permission from Das et al. (2021) @ Elsevier [23]).

Variables	Value
pH	7.8
Phenol, mg/L	15
BOD, mg/L	105
COD, mg/L	750
Oil & grease, mg/L	25
Iron, mg/L	6

CDs below 200 A/m^2 were insufficient to reduce all contaminants below acceptable limits. Consequently, a CD of 200 A/m^2 and an EC treatment period of 30 minutes were determined as the optimal parameters for the experiment, and beyond these conditions, the removal of the contaminants resulted in negligible. For this study, they chose Al as an electrode for performing the EC investigations. It was noted that using Fe electrodes during EC led to color formation in the treated samples after each experiment. Alternatively, Al electrodes were used to stop color from forming in the final treated water [25]. In the EC method, the solution's pH gradually rises from 7.8 until it reaches an equilibrium state with a pH of 8.5. This equilibrium was achieved due to the buffering capacity of Al(OH)3 formed through the experiment. The pH increase resulted from the generation of OH^- ions at the cathode surface and the evolution of H_2 near the cathode. Furthermore, as the CD and electrolysis time increased, a greater amount of ions and H_2 were produced in the experiment. Consequently, the solution's pH increased as the EC process progressed. It was examined that as the CD is increased from 100 to 200 A/m^2, there is a corresponding increase in the phenol removal efficacy of 70% to 98% in 30 minutes of the treatment process. The EC process revealed that the efficacy of phenol reduction improved with increasing electrolysis time when Al electrodes were used. Furthermore, the *in situ* formation of aluminum hydroxide flocs is intensified by the higher CD, leading to increased floc precipitation. This enhancement contributes to the reduction in phenol concentration. Furthermore, a reduction in Fe concentration from 15 mg/L to 0.3 mg/L was observed. Similarly, under operating conditions of a CD of 100 to 200 mA/cm^2 and a treatment time of 30 minutes, they found that the BOD reduction was 85.7%, and COD, iron, and oil and grease reduction efficiencies increased from 58.5 to 90.5%, 65 to 97.5%, and 40 to 88%, respectively [23].

8.3.4 Fluoride and Arsenic Mitigation in Drinking Water by the Electrocoagulation Process

Drinking water contaminated with fluoride and arsenic may possess detrimental effects on human well-being. Arsenic is an element that occurs naturally and can enter into water supplies due to industrial or geological processes. Long-term ingestion of arsenic in drinking water has been connected to several health concerns, such as skin lesions, lung and bladder tumors, cardiovascular disorders, and issues with breathing. Arsenicosis may result from long-term consumption of water tainted with arsenic. While fluoride is beneficial in small amounts for dental health, it can pose health risks when present in excess in drinking water. High fluoride levels can cause dental and skeletal fluorosis, leading to mottled teeth, skeletal deformities, and weakened bones. Prolonged exposure to elevated fluoride levels may result in joint stiffness, pain, and increased fracture risk, particularly in areas where naturally occurring fluoride content is high in water sources [26]. For this study, the well water was sourced from Guadiana Valley, Durango State, Mexico, which had coexisting levels of arsenic and fluoride, as shown in Table 8.4.

The EC system consisted of a rectangular acrylic reactor of dimensions 22 cm x 12 cm x 12 cm, with an effective capacity of 1.5 L. The electrodes were made of Fe and Al perforated plates, with a 0.5 cm gap between them. The surfaces of the Fe and Al electrodes were cleaned by immersing them in a 1M HCl and a 20% NaOH solution, respectively, and then washing them with distilled water. For every experiment, the plates were rinsed with ethyl alcohol to get rid of any oily residue. To produce the greatest amount of coagulant species for each of the pollutants, the Fe electrode was employed for arsenic species and the Al electrode for fluoride species. The EC experiment was conducted using four electrodes, including two Fe electrodes (one was the anode and the other was the cathode) and two Al electrodes (one was the anode and the other was the cathode). A monopolar connection was made in parallel between the electrodes and a DC supply. The electrodes had an active area of 577.2 cm^2. A magnetic stirrer was used to swirl the mixture continuously at 150 rpm during the trials. The pH, CD, and treatment time are the three most important variables in the EC method which influence the performance of the process. Under ideal working conditions of a pH of 5, a treatment time of 15 minutes and a CD of 4.5 mA/cm^2, the reduction yield for fluoride was 85.68% and for arsenic roughly 100%. This led to final concentrations significantly below WHO recommendations and national

TABLE 8.4
The Properties of Well Water Sourced from Guadiana Valley, Mexico (Reproduced with permission from López-Guzmán et al. (2019) @ Elsevier [27]).

Parameters	Value
pH	7.7
Conductivity, μS/cm	409.9
Fluoride, mg/L	4.18
Total arsenic, mg/L	24.48
Nitrate, mg/L	14.07
Chloride, mg/L	18.43

standards. The elimination of arsenic was not significantly impacted by the initial pH adjustment. However, the acidic pH conditions help to remove fluoride, with an optimal removal rate of 94.98% to attain 1.0 mg/L of fluoride in the water as the final concentration [27].

8.3.5 Treatment of Drinking Water Source Contaminated with Viruses and Bacteria by the Electrocoagulation Process

Many different types of microorganisms, including bacteria, viruses, protozoa, and more, can be found in water and wastewater. Their presence in water can result in waterborne illnesses, which are extremely dangerous to people's health. Disinfection is regarded as one of the multiple-barrier water purification processes that ensures the biological safety of drinking water. For more than a century, chlorination has been widely employed as an economical and effective disinfection technique, although it has several clear drawbacks, including the production of hazardous by-products such as THMs. Recently, electrochemical disinfection has shown promise as a viable water treatment solution. It has been shown that EC is a potential approach for eliminating a variety of microorganisms from various water systems, such as drinking water or repurposed water, ranging from bacteria to viruses [28]. According to the majority of research, Escherichia coli (E. coli) is a more particular microbiological pathogen indicator of disinfection [29].

For this study, urban effluent was collected from a WWTP in Reus, Spain. On various days, multiple sets of new samples were taken from the secondary treatment effluent (activated sludge), and the primary treatment effluent. These wastewaters were chosen to evaluate the impact of varying organic matter contents on the reduction of microorganisms in the EC. They were put into sterile containers, moved to a larger isothermal container at 4 °C, and then microbiologically characterized within 24 to 48 hours of collection. Finally, they were stored at 4 °C. The tests found that spores, protozoa, bacteriophages, and bacteria were present. This study was performed using an EC reactor having Fe/Fe electrodes at a CD of 200 A/m^2 and an electrolysis time of 90 minutes. The anode and the cathode comprised Fe plates each with an area of 10 cm^2, spaced apart by 1 cm. The pH value increases from its initial pH of 7.54 to 9.08, but the conductivity barely altered, fluctuating between 2.50 and 2.04 mS/cm. There was an unequal disappearance of various bacteria during the treatment of both samples. The least durable species were the somatic coliphages, which vanished completely from both matrices after 20 minutes and were followed by the eukaryotes, whose concentration sharply dropped at that point. Furthermore, the concentration of heterotrophic bacteria reduced to around 103 CFU/mL in 90 minutes, meaning that the primary and secondary effluents had reductions of 3.66 and 2.81 log units, respectively [30].

8.3.6 Hybrid Adsorption–Electrocoagulation Process for the Treatment of Petroleum Industry Wastewater

The hybrid adsorption–electrocoagulation process offers several advantages for treating petroleum industry wastewater. This hybrid system combines the benefits of both adsorption and EC techniques, leading to enhanced removal efficiencies of various pollutants present in petroleum wastewater. Moreover, the hybrid approach can be tailored and optimized to target specific pollutants present in petroleum wastewater, leading to efficient treatment outcomes. It also offers flexibility in terms of scalability and operational parameters, making it suitable for both small-scale and large-scale wastewater treatment plants in the petroleum industry.

The hybrid adsorption–electrocoagulation process often requires lower energy consumption than conventional treatment methods, thereby contributing to cost savings and environmental sustainability. For this case study, the effluent was sourced from the BARAKI refinery plant situated in Alger, Algeria.

A total of 20 L of wastewater was collected after the DAF unit, before starting the biological treatment process. The properties of these samples, which were thereafter kept at 4 °C until needed, are listed in Table 8.5. The effluent's characteristics show a substantial degree of contamination, which can be harmful to the microorganisms involved in the ensuing biological treatment procedure. Furthermore, the effluent demonstrates a low electrical conductivity of 2.45 mS/cm. In the literature, researchers have explored various methods for treating petroleum wastewater. It has been generally observed that employing a combination or hybridization of different methods under suitable conditions is more effective than using a single method alone. For instance, Nigri et al. (2020) studied the process of treating real petroleum effluent to remove organic chemicals, calcium, and strontium using simultaneous adsorption and EC method. They noted removal efficiencies of 72%, 88%, and 52% for strontium, calcium, and COD, respectively [32]. The objective of this case study was to evaluate the efficiency of EC and adsorption techniques in eliminating COD, turbidity, and TSS. Furthermore, under the optimum working conditions of each technique, the potential of combining these two methods in a hybrid approach was investigated to enhance removal efficacy. The hybrid process was conducted using suboptimal factors such as an adsorbent dose of 0.5 g/L and a temperature of 25°C for the adsorption process and a CD of 4 mA/cm^2 and conductivity of 4 mS/cm for the EC technique. The EC process proved to be more effective than adsorption. This difference in effectiveness could be attributed to the EC process's ability to remove

TABLE 8.5
The properties of petroleum industry wastewater (Reproduced with permission from Chabani et al. (2024) @ Elsevier [31]).

Parameter	Value
pH	9
Turbidity, NTU	50–57
COD, mg/L	350–360
TSS, mg/L	147–150
BOD_5, mg/L	104
Electrical conductivity, mS/cm	2.45

both inorganic and organic impurities by adsorbing them into the generated aluminum hydroxide complexes. On the other hand, the main function of adsorption is to move organic contaminants from one stage to another and to fill the pores of activated carbon with the suspension. To eliminate COD from refinery wastewater, they examined the effects of hybrid EC and adsorption utilizing powdered activated carbon (PAC). They evaluated this strategy against the effectiveness of each technique separately to reuse the treated effluent for uses such as irrigation or reintegrate it into refinery processes. Their findings revealed a synergistic effect between EC and the adsorption process during the treatment, particularly under less favorable conditions. This combined method proved to be a useful and efficient solution, achieving a reduction of 82% in COD, 98.64% in total suspended solids (TSS), and 96.5% in turbidity from the effluent. In comparison to conventional EC techniques, these outcomes were obtained with a shorter period of treatment utilizing a PAC dose of 0.5 g/L and a CD of 4 mA/cm^2. Applying this combined treatment not only offers a chance to treat Algiers refinery effluent on a greater scale but also permits the subsequent use of purified wastewater within refinery processes. This offers a viable solution to improve environmental sustainability and drastically reduce freshwater use in refinery operations [31].

8.4 COMPARISON OF ELECTROCOAGULATION TREATMENT WITH CONVENTIONAL PROCESSES

Comparison of EC treatment with conventional processes involves evaluating various aspects such as efficiency, cost-effectiveness, environmental impact, and applicability across different industries. EC is a relatively newer technology that utilizes electrical energy to remove contaminants from water, while conventional processes encompass a range of established methods such as sedimentation, filtration, chemical coagulation, and biological treatment. Each approach has its strengths and weaknesses, and a comprehensive comparison can help identify the most suitable method for specific water treatment challenges. In terms of efficiency, EC offers several advantages over conventional processes. EC is highly effective in removing a wide range of contaminants, including suspended solids, organic compounds, heavy metals, and pathogens, due to the electrochemical reactions that occur during treatment. The process destabilizes pollutants, facilitating their agglomeration and precipitation, which can result in faster and more thorough removal than conventional methods. Moreover, EC can operate over a wide range of pH levels and does not require the addition of chemicals in many cases, making it versatile and adaptable to various water quality conditions [33].

However, while EC excels in certain aspects of efficiency, conventional processes also have their strengths. For example, sedimentation and filtration are well-established methods for removing large particles and solids from water, and they are often more straightforward to implement and maintain than EC systems. Chemical coagulation, another conventional method, can be highly effective for specific contaminants and is often preferred for its simplicity and ease of use in certain applications. Biological treatment processes, such as activated sludge and biofiltration, offer advantages for organic pollutant removal and nutrient removal but may be less effective for certain types of contaminants like heavy metals. Cost-effectiveness is another critical factor for comparing EC with conventional processes. While the initial capital investment for EC systems may be higher than some conventional methods, EC can offer long-term cost savings due to lower chemical and energy consumption, reduced sludge production, and potentially higher treatment efficiency, leading to lower operating costs over time. Additionally, EC systems are often modular and scalable, allowing for easier expansion or adjustment to changing treatment needs compared to conventional treatment plants, which may require significant infrastructure upgrades or modifications [34].

Environmental impact is also an important consideration in comparing water treatment processes. EC typically generates less sludge than conventional chemical coagulation processes, thereby reducing disposal costs and minimizing the environmental footprint of treatment operations. In addition, EC systems can be designed to operate with renewable energy sources, further reducing their carbon footprint and overall environmental impact. However, both EC and conventional processes may have potential environmental drawbacks depending on factors such as chemical usage, energy consumption, and the disposal or reuse of treated water and sludge. Ultimately, the choice between EC and conventional treatment methods requires careful consideration of these factors to determine the most suitable approach for achieving optimal water quality and treatment efficiency [35].

8.5 LIMITATIONS AND FUTURE PERSPECTIVES

The effectiveness of the EC process in eliminating contaminants from wastewater has been extensively demonstrated in both laboratory- and pilot-scale plants, as evidenced by numerous studies. However, further research is needed to validate the operational parameters utilized to scale up the process, but these parameters can change based on the kind of effluents. Despite the strong dependence of pollutant removal mechanisms on physicochemical properties, the EC method is generally considered extremely proficient compared to conventional processes. Enhanced contaminant removal efficacy is usually obtained when using Al or Fe electrodes, but complete elimination is not always possible. It has also been documented that other electrodes such as Zn and Cu can be used as sacrificial anodes in the EC method. The efficacy of the EC method depends on numerous operational factors, including solution pH, CD, electrode type, supporting electrolyte, inter-electrode spacing, and electrolysis period. The treatment performance of the EC method can be greatly increased by optimizing reactor designs, and electrode connection choices, and using a simulation model in conjunction with cutting-edge electrode materials. Furthermore, integrating the EC method with other water treatment approaches improves pollutant removal efficiency. Hybrid EC processes notably reduce the overall energy consumption and operational costs.

8.6 CONCLUSION

The increasing pace of industrialization has brought numerous challenges related to water pollution to the forefront. In addressing these challenges, EC technology presents extensive prospects for application. This chapter focuses on the recent advancements in EC reactor design and configuration. The design of EC reactors still heavily relies on experience and trial-and-error methods, lacking systematic design principles and optimization techniques. To address this issue, mathematical models and computational software such as computational fluid dynamics (CFD) tools can offer valuable support for reactor design. Commonly used mathematical models like the Navier–Stokes (NS) equation and Nernst–Planck (NP) equation are frequently employed in simulating EC reactors. The NS equation simulates fluid momentum transfer within the EC reactor, while the NP equation describes the mass transfer of ionic species into the solution. CFD is a crucial tool for studying various phenomena such as multiphase flow, momentum transfer, heat and mass transfer, and chemical reactions, integrating fluid mechanics with computer science. In EC research, CFD simulation has attracted increasing attention from researchers in modern years. The important EC parameters are electrode geometry, fluid flow patterns, current distribution, and economic analyses, which can be optimized using CFD simulation.

REFERENCES

[1] D.B. Miklos, C. Remy, M. Jekel, K.G. Linden, J.E. Drewes, U. Hübner, Evaluation of advanced oxidation processes for water and wastewater treatment—a critical review, *Water Res.* 139 (2018) 118–131. https://doi.org/10.1016/j.watres.2018.03.042.

[2] B.S. Yildiz, Chapter 18—Water and wastewater treatment: Biological processes, in: F. Zeman (Ed.), *Metropolitan Sustainability*, Woodhead Publishing, 2012: pp. 406–428. https://doi.org/10.1533/9780857096463.3.406.

[3] R. Singh, M.K. Purkait, Chapter 4—Microfiltration membranes, in: A.F. Ismail, M.A. Rahman, M.H.D. Othman, T. Matsuura (Eds.), *Membrane Separation Principles and Applications*, Elsevier, 2019: pp. 111–146. https://doi.org/10.1016/B978-0-12-812815-2.00004-1.

[4] M. Bharti, P.P. Das, M.K. Purkait, A review on the treatment of water and wastewater by electrocoagulation process: Advances and emerging applications, *J. Environ. Chem. Eng.* 11 (2023) 111558. https://doi.org/10.1016/j.jece.2023.111558.

[5] P.P. Das, A.D. Sontakke, M.K. Purkait, Chapter 2—Electrocoagulation process for wastewater treatment: Applications, challenges, and prospects, in: M.P. Shah (Ed.), *Development in Wastewater Treatment Research and Processes*, Elsevier, 2023: pp. 23–48. https://doi.org/10.1016/B978-0-323-95684-0.00015-4.

[6] P.P. Das, M. Sharma, M.K. Purkait, Recent progress on electrocoagulation process for wastewater treatment: A review, *Sep. Purif. Technol.* 292 (2022) 121058. https://doi.org/10.1016/j.seppur.2022.121058.

[7] Y.O.A. Fouad, A.H. Konsowa, H.A. Farag, G.H. Sedahmed, Performance of an electrocoagulation cell with horizontally oriented electrodes in oil separation compared to a cell with vertical electrodes, *Chem. Eng. J.* 145 (2009) 436–440. https://doi.org/10.1016/j.cej.2008.04.027.

[8] I. Hassan, I. Nirdosh, G.H. Sedahmed, Separation of oil from oil-water emulsions by electrocoagulation in an electrochemical reactor with a fixed-bed anode, *Water Air Soil Pollut.* 226 (2015). https://doi.org/10.1007/s11270-015-2521-4.

[9] A. Genc, B. Bakirci, Treatment of emulsified oils by electrocoagulation: Pulsed voltage applications, *Water Sci. Technol.* 71 (2015) 1196–1202. https://doi.org/10.2166/wst.2015.092.

[10] P. Maha Lakshmi, P. Sivashanmugam, Treatment of oil tanning effluent by electrocoagulation: Influence of ultrasound and hybrid electrode on COD removal, *Sep. Purif. Technol.* 116 (2013) 378–384. https://doi.org/10.1016/j.seppur.2013.05.026.

[11] C. An, G. Huang, Y. Yao, S. Zhao, Emerging usage of electrocoagulation technology for oil removal from wastewater: A review, *Sci. Total Environ.* 579 (2017) 537–556. https://doi.org/10.1016/j.scitotenv.2016.11.062.
[12] S. Zhao, G. Huang, G. Cheng, Y. Wang, H. Fu, Hardness, COD and turbidity removals from produced water by electrocoagulation pretreatment prior to reverse osmosis membranes, *Desalination* 344 (2014) 454–462. https://doi.org/10.1016/j.desal.2014.04.014.
[13] K. Ulucan, H.A. Kabuk, F. Ilhan, U. Kurt, Electrocoagulation process application in bilge water treatment using response surface methodology, *Int. J. Electrochem. Sci.* 9 (2014) 2316–2326. https://doi.org/10.1016/s1452-3981(23)07929-4.
[14] O. Chavalparit, M. Ongwandee, Optimizing electrocoagulation process for the treatment of biodiesel wastewater using response surface methodology, *J. Environ. Sci.* 21 (2009) 1491–1496. https://doi.org/10.1016/S1001-0742(08)62445-6.
[15] M.A. Bustamante, C. Paredes, R. Moral, J. Moreno-Caselles, A. Pérez-Espinosa, M.D. Pérez-Murcia, Uses of winery and distillery effluents in agriculture: Characterisation of nutrient and hazardous components, *Water Sci. Technol.* 51 (2005) 145–151. https://doi.org/10.2166/wst.2005.0018.
[16] P.P. Das, M.K. Purkait, Chapter 5—Removal of industrial wastewater contaminants by ozonation process, in: M.P. Shah (Ed.), *Development in Wastewater Treatment Research and Processes*, Elsevier, 2024: pp. 75–89. https://doi.org/10.1016/B978-0-443-13884-3.00025-1.
[17] Y. Yavuz, EC and EF processes for the treatment of alcohol distillery wastewater, *Sep. Purif. Technol.* 53 (2007) 135–140. https://doi.org/10.1016/j.seppur.2006.08.022.
[18] R. Davarnejad, M. Nikseresht, I. Ajideh, An efficient technique for dairy wastewater treatment, *Int. J. Dairy Technol.* 71 (2018) 532–538. https://doi.org/10.1111/1471-0307.12475.
[19] E. Gengec, M. Kobya, E. Demirbas, A. Akyol, K. Oktor, Optimization of baker's yeast wastewater using response surface methodology by electrocoagulation, *Desalination* 286 (2012) 200–209. https://doi.org/10.1016/j.desal.2011.11.023.
[20] M. Eyvaz, E. Gürbulak, S. Kara, E. Yüksel, Preventing of cathode passivation/deposition in electrochemical treatment methods—a case study on winery wastewater with electrocoagulation, in: M. Aliofkhazraei (Ed.), *Modern Electrochemical Methods in Nano, Surface and Corrosion Science*, IntechOpen, Rijeka, 2014: p. Ch. 8. https://doi.org/10.5772/58580.
[21] P. Drogui, M. Asselin, S.K. Brar, H. Benmoussa, J.-F. Blais, Electrochemical removal of pollutants from agro-industry wastewaters, *Sep. Purif. Technol.* 61 (2008) 301–310. https://doi.org/10.1016/j.seppur.2007.10.013.
[22] B. Merzouk, K. Madani, A. Sekki, Using electrocoagulation—electroflotation technology to treat synthetic solution and textile wastewater, two case studies, *Desalination* 250 (2010) 573–577. https://doi.org/10.1016/j.desal.2009.09.026.
[23] P.P. Das, Anweshan, M.K. Purkait, Treatment of cold rolling mill (CRM) effluent of steel industry, *Sep. Purif. Technol.* 274 (2021) 119083. https://doi.org/10.1016/j.seppur.2021.119083.
[24] P.P. Das, P. Mondal, Anweshan, A. Sinha, P. Biswas, S. Sarkar, M.K. Purkait, Treatment of steel plant generated biological oxidation treated (BOT) wastewater by hybrid process, *Sep. Purif. Technol.* 258 (2021) 118013. https://doi.org/10.1016/j.seppur.2020.118013.
[25] M. Changmai, M. Pasawan, M.K. Purkait, Treatment of oily wastewater from drilling site using electrocoagulation followed by microfiltration, *Sep. Purif. Technol.* 210 (2019) 463–472. https://doi.org/10.1016/j.seppur.2018.08.007.
[26] P. Mondal, A. Chattopadhyay, Environmental exposure of arsenic and fluoride and their combined toxicity: A recent update, *J. Appl. Toxicol.* 40 (2020) 552–566. https://doi.org/10.1002/jat.3931.

[27] M. López-Guzmán, M.T. Alarcón-Herrera, J.R. Irigoyen-Campuzano, L.A. Torres-Castañón, L. Reynoso-Cuevas, Simultaneous removal of fluoride and arsenic from well water by electrocoagulation, *Sci. Total Environ.* 678 (2019) 181–187. https://doi.org/10.1016/j.scitotenv.2019.04.400.

[28] B. Bicudo, D. van Halem, S.A. Trikannad, G. Ferrero, G. Medema, Low voltage iron electrocoagulation as a tertiary treatment of municipal wastewater: Removal of enteric pathogen indicators and antibiotic-resistant bacteria, *Water Res.* 188 (2021) 116500. https://doi.org/10.1016/j.watres.2020.116500.

[29] G. Sruthi, M.M. Ahammed, A.R. Makwana, Effect of source water/wastewater quality on bacterial removal during electrocoagulation, *Water Sci. Technol.* 77 (2018) 1460–1468. https://doi.org/10.2166/wst.2018.024.

[30] E. Anfruns-Estrada, C. Bruguera-Casamada, H. Salvadó, E. Brillas, I. Sirés, R.M. Araujo, Inactivation of microbiota from urban wastewater by single and sequential electrocoagulation and electro-Fenton treatments, *Water Res.* 126 (2017) 450–459. https://doi.org/10.1016/j.watres.2017.09.056.

[31] M. Nor El houda, M. Chabani, S. Bouafia-Chergui, A. Touil, Removal of chemical oxygen demand from real petroleum refinery wastewater through a hybrid approach: Electrocoagulation and adsorption, *Chem. Eng. Process. Process Intensif.* 196 (2024) 109680. https://doi.org/10.1016/j.cep.2024.109680.

[32] E.M. Nigri, A.L.A. Santos, S.D.F. Rocha, Removal of organic compounds, calcium and strontium from petroleum industry effluent by simultaneous electrocoagulation and adsorption, *J. Water Process Eng.* 37 (2020) 101442. https://doi.org/10.1016/j.jwpe.2020.101442.

[33] P. Cañizares, F. Martínez, C. Jiménez, C. Sáez, M.A. Rodrigo, Technical and economic comparison of conventional and electrochemical coagulation processes, *J. Chem. Technol. Biotechnol.* 84 (2009) 702–710. https://doi.org/10.1002/jctb.2102.

[34] F. Akbal, S. Camcı, Comparison of electrocoagulation and chemical coagulation for heavy metal removal, *Chem. Eng. Technol.* 33 (2010) 1655–1664. https://doi.org/10.1002/ceat.201000091.

[35] D.-G. Kim, R.J.S. Palacios, S.-O. Ko, Characterization of sludge generated by electrocoagulation for the removal of heavy metals, *Desalin. Water Treat.* 52 (2014) 909–919. https://doi.org/10.1080/19443994.2013.826776.

9 Environmental and Economic Sustainability of the EC Process

9.1 INTRODUCTION

Electrocoagulation plays a crucial role in promoting environmental sustainability by effectively removing various pollutants from water sources. Through the application of an electrical current, metal ions are generated, leading to coagulation and precipitation of contaminants such as heavy metals, suspended solids, organic compounds, and pathogens. Unlike conventional treatment methods, EC does not rely on the addition of chemical coagulants, thus reducing the generation of harmful by-products and minimizing the environmental footprint of the treatment process. Moreover, the ecological sustainability of EC depends on factors such as system design, operational practices, and site-specific conditions. Integration with other treatment technologies, such as membrane filtration [1] or advanced oxidation processes, can enhance the overall efficiency and environmental performance of EC systems, enabling tailored solutions for diverse wastewater treatment challenges. Furthermore, pilot-scale studies and field demonstrations play a crucial role in validating the environmental benefits of EC under real-world conditions, providing valuable data for decision-making and technology adoption. In addition to its environmental benefits, EC holds immense potential for achieving economic sustainability by offering cost-effective wastewater treatment solutions.

The primary advantages of EC lie in its operational efficiency and scalability, enabling the treatment of large volumes of water with minimal energy consumption and maintenance requirements. EC offers a more economically viable alternative, particularly for industries and communities facing budget constraints, than conventional treatment methods, which often entail significant capital and operational costs. The by-products generated during EC, such as sludge rich in metal hydroxides, can be valorized through processes such as metal recovery or agricultural applications, creating additional revenue streams and enhancing the overall economic viability of the treatment process [2]. Additionally, EC offers economic benefits through its potential for resource recovery and reuse. By effectively removing pollutants from wastewater streams, EC generates treated water that can be recycled for various purposes, including irrigation, industrial processes, and potable water supply. This not only reduces the demand for freshwater resources but also lowers operational costs associated with water procurement and disposal [3].

However, this chapter aims at exploring the benefits and challenges associated with the EC process, focusing on its application in wastewater treatment. It also highlights

DOI: 10.1201/9781003499671-9

the economic advantages of the EC process, such as lower operational costs and scalability. In addition to environmental considerations, sludge dewatering operations also have economic implications for wastewater treatment facilities. Sludge dewatering is a critical component of wastewater treatment operations, enabling the efficient management and disposal of sludge generated during the treatment process. However, this chapter provides a detailed exploration of EC and its contributions to environmental and economic sustainability in wastewater treatment.

9.2 ENVIRONMENTAL ASSESSMENT OF THE ELECTROCOAGULATION PROCESS

Environmental assessment of the EC process involves evaluating its impact on various aspects of the environment, including water quality, energy consumption, greenhouse gas emissions, sludge generation, and overall sustainability. EC has emerged as a promising technology for wastewater treatment, offering several advantages over conventional processes. However, a thorough environmental assessment is essential to understand its potential benefits and limitations in different contexts [4]. Firstly, EC can significantly improve water quality by effectively removing various pollutants, including suspended solids, organic compounds, heavy metals, and pathogens, from wastewater streams. Unlike conventional processes such as chemical coagulation, EC does not typically require the addition of chemical coagulants, thereby reducing the risk of introducing additional contaminants into the environment. Additionally, EC can achieve high removal efficiencies even in complex wastewater matrices, contributing to the protection of receiving water bodies and ecosystems. Moreover, EC has the potential to reduce energy consumption compared to some conventional treatment methods. While EC systems require electricity to operate, advancements in electrode materials, reactor design, and process optimization have led to increased energy efficiency over time. By optimizing operating conditions and utilizing renewable energy sources, such as solar or wind power, the environmental footprint of EC can be further minimized, enhancing its overall sustainability [5]. The generation and management of sludge are important considerations in environmental assessments of wastewater treatment technologies. EC can produce sludge through the precipitation and aggregation of contaminants, which may require proper handling and disposal to prevent environmental impacts. However, the characteristics of EC-generated sludge differ from those produced by chemical coagulation, often exhibiting better dewaterability and reduced hazardousness. Innovative approaches such as sludge recycling, resource recovery, or beneficial reuse can further mitigate the environmental impact of EC-generated sludge, turning it into a valuable resource rather than a waste stream [6].

9.2.1 Sludge Dewatering Operation

Sludge dewatering is a critical operation in wastewater treatment plants aimed at reducing the volume and moisture content of sludge, making it easier and more cost-effective to handle, transport, and dispose of. This process involves the removal of water from sludge solids, transforming the slurry-like material into a drier, more

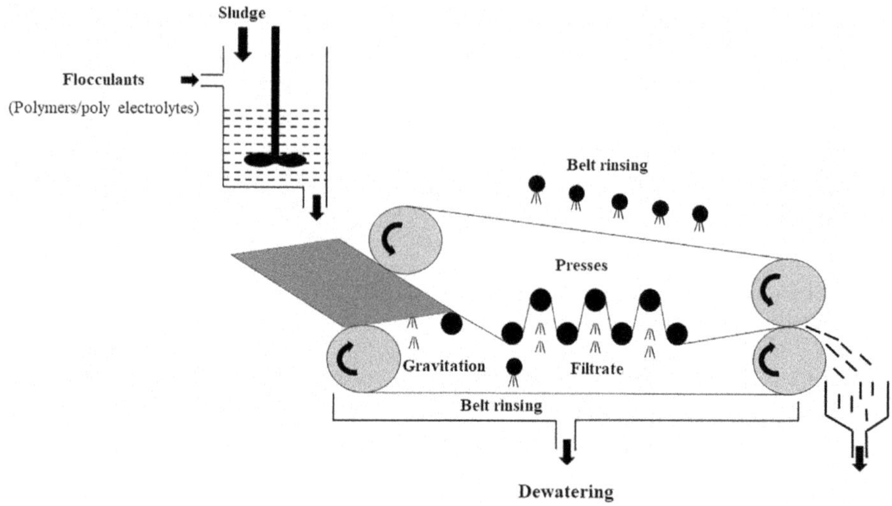

FIGURE 9.1 Schematic diagram of dewatering of sludge using a belt filter press.

manageable form. Various techniques and technologies are employed for sludge dewatering [7], each with its advantages, limitations, and environmental considerations. One commonly used method for sludge dewatering is mechanical dewatering, which typically involves the use of equipment such as filter presses, belt filter presses, centrifuges, and screw presses, as shown in Figure 9.1. These mechanical devices apply pressure, vacuum, or centrifugal force to separate water from sludge solids. Filter presses, for example, utilize a series of filter plates and frames to compress sludge between them, forcing water to drain out through filter cloths.

Centrifuges utilize centrifugal force to separate solids and liquids, while screw presses employ a rotating screw to squeeze water out of the sludge [8]. Another commonly used dewatering technique is thermal drying, which involves applying heat to sludge to evaporate water and produce a dry solid material. This process typically requires specialized equipment such as rotary dryers, fluidized bed dryers, or belt dryers. Thermal drying can be energy-intensive and may require additional treatment to handle off-gas emissions generated during the drying process. However, it can produce a relatively dry and stable product suitable for further use or disposal. Environmental considerations play a significant role in sludge dewatering operations, particularly regarding the handling and disposal of dewatered sludge. Depending on its composition and treatment history, dewatered sludge may be suitable for beneficial reuse in applications such as land application, soil amendment, or composting. Therefore, care must be taken to ensure that dewatered sludge meets regulatory requirements and does not pose environmental or public health risks. In some cases, dewatered sludge may require further treatment or stabilization before it can be safely reused or disposed of [9]. In addition to environmental considerations, sludge dewatering operations also have economic implications for wastewater treatment facilities. The cost of dewatering can vary depending on factors such as the volume and composition of sludge, the dewatering technology employed, energy consumption, chemical usage, and disposal costs. Facilities must carefully evaluate the costs and benefits of different dewatering options to optimize their operations and achieve regulatory compliance while minimizing costs. Overall, sludge dewatering is a critical component of wastewater treatment operations, enabling the efficient management and disposal of sludge generated during treatment [10].

9.2.2 Reduction in Carbon Footprint

A carbon footprint quantifies the greenhouse gases generated from burning fossil fuels for various purposes like heating, electricity, and transportation. This measurement is typically stated in tones (or kilograms) of carbon dioxide equivalent [11]. For instance, clothing contributes to carbon emissions at every stage of a textile product's life span and is one of the basic needs of humans. Therefore, to preserve our world, everyone needs to reduce carbon footprints. This is why reducing carbon footprints is such a critical topic nowadays. The clothing and textiles industry has a long and intricate supply chain, which contributes significantly to the development of carbon footprints and is one of the main sources of greenhouse gas emissions. The carbon footprint is threatened by the manufacture, distribution, utilization, consumption, and disposal of textile items [12]. There are numerous approaches to lessen the carbon footprint of textile products over their whole life span: recycling process waste rather than throwing it in the trash and recycling. In contrast to the previous decade, human beings have started encouraging recycling activities in the current decade due to the dangerous environmental consequences as well as other factors like governmental legislation and incentives in the form of monetary advantages offered to consumers when they return products for recycling. One effective and promising strategy to lessen carbon footprint is recycling. Textile products can be recycled in a variety of ways to lower their carbon footprint. This chapter provides a detailed explanation of the possible benefits of recycling textile process waste and textile goods at the final stage of their life, particularly reduction of carbon footprint.

Figure 9.2 illustrates a conceptual framework for the idea of reducing carbon footprints through recycling [13]. From raw materials to finished goods, a textile product must go through numerous procedures in the supply chain, each of which has an impact on carbon emissions. Throughout the lifespan of a good, waste textile materials in various forms occur at various times. Pre- and post-consumer textile waste are two categories into which it can fall. Textile items that are thrown after they become outdated and unusable are classified as post-consumer waste, while pre-consumer waste can arise at any stage of the production line in the textile and garment industry. During various stages of a product's life span, recycling textile waste will undoubtedly have an effect on lowering its carbon footprint [14].

Reducing the carbon footprint using EC represents a significant opportunity for sustainable wastewater treatment practices. The application of EC offers several pathways to mitigate carbon emissions compared to conventional treatment methods, thereby contributing to environmental preservation and climate change mitigation efforts. This paragraph explores how EC can lead to a reduction in the carbon

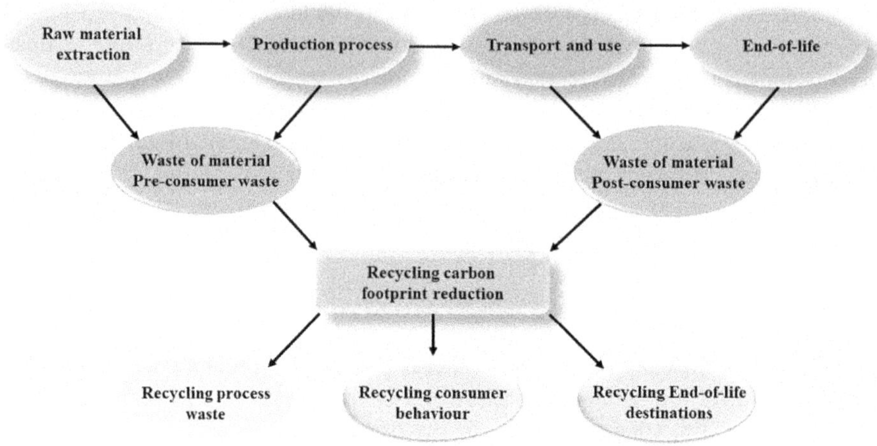

FIGURE 9.2 Framework of carbon footprint reduction through recycling.

footprint across various aspects of wastewater treatment. Firstly, EC presents a more environmentally friendly alternative to conventional treatment methods by minimizing the use of chemical coagulants. In traditional coagulation processes, such as chemical precipitation, large quantities of chemical coagulants are often required to achieve adequate pollutant removal. These chemicals, typically aluminum or iron salts, contribute to carbon emissions through their production, transportation, and application [15]. In contrast, EC relies on electrochemical reactions to destabilize contaminants, eliminating the need for chemical additives in many cases. By reducing reliance on chemical coagulants, EC helps decrease carbon emissions associated with chemical manufacturing and transportation, thus lowering the overall carbon footprint of wastewater treatment operations.

Furthermore, EC can lead to energy savings compared to some conventional treatment methods, further reducing carbon emissions. While EC systems require electricity to power the electrodes and facilitate the electrochemical reactions, advancements in technology have led to improvements in energy efficiency over time. The use of renewable energy sources [16], such as solar or wind power, to supply electricity to EC systems can further reduce carbon emissions associated with energy generation. By harnessing clean and sustainable energy sources, wastewater treatment facilities can significantly lower their carbon footprint while achieving effective pollutant removal through EC. Moreover, the by-products generated during the EC process can offer opportunities for carbon sequestration or utilization, contributing to the overall greenhouse gas mitigation efforts. The sludge produced in EC systems contains precipitated contaminants, which can be rich in carbonaceous materials depending on the composition of the wastewater. By properly managing and treating EC-generated sludge, facilities can potentially sequester carbon in stabilized forms, reducing the release of greenhouse gases into the atmosphere [13].

9.2.3 Life Cycle Assessment

The term "life cycle" describes a product's whole existence, including the extraction of resources, manufacturing, use, and eventual disposal. It is reflected in the cycle aspect that resources taken out of the environment are eventually put back into it.

Analysis and evaluation of product life cycles are referred to as life cycle assessment or LCA.

LCA is mainly divided into four stages, as shown in Figure 9.3.

(1) **Raw material acquisition:** This phase includes the extraction and processing of materials required for the EC system, chemicals (which support the electrolytes such as NaCl), and electrodes (which are usually made of Fe or Al).
(2) **Electrocoagulation treatment:** The electrocoagulation process takes place during this operational step. It covers the energy used to operate the unit, the input of wastewater, and the production of sludge or flocs that hold the pollutants that have been removed.
(3) **Sludge management:** This phase deals with the treatment of the sludge that results from the EC process. The possibilities include dumping the sludge in landfills, applying it to land after treatment (like composting), or perhaps even extracting valuable materials from it.
(4) **End-of-life disposal:** This phase involves the decommissioning of the electrocoagulation unit once it reaches the end of its operational life. This encompasses activities such as disassembly, recycling, or disposal of the unit's components.

LCA studies can provide valuable insights into the environmental performance of EC compared to conventional treatment methods, taking into account the entire

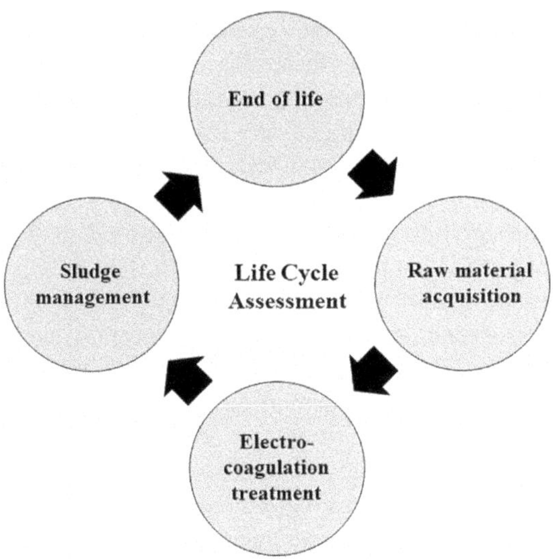

FIGURE 9.3 Life cycle assessment for the electrocoagulation process.

suitability of treated water for reuse depends on factors such as the level of contaminant removal achieved, the presence of residual by-products, and the specific requirements of the intended reuse application. For example, treated water intended for irrigation may need to meet certain nutrient and microbiological standards, while water destined for industrial processes may require additional treatment to remove specific pollutants or meet quality specifications [20].

9.3 ECONOMIC ASSESSMENT OF THE EC PROCESS

Economic assessment of the EC process is crucial for evaluating its viability and competitiveness as a wastewater treatment technology. This assessment involves analyzing various economic aspects, including capital and operational costs, cost-effectiveness, return on investment (ROI), and overall economic feasibility. By examining these factors, stakeholders can make informed decisions regarding the adoption, optimization, and implementation of EC systems. This paragraph explores the key components of economic assessment for EC, highlighting its potential benefits, challenges, and strategies for cost optimization [21].

The primary consideration in the economic assessment of EC is the initial capital investment required to install and commission EC systems. Capital costs for EC equipment typically include the purchase of electrodes, reactors, power supplies, control systems, and auxiliary equipment. The size and complexity of EC systems depend on factors such as the scale of wastewater treatment, flow rates, contaminant concentrations, and treatment objectives. Additionally, site-specific factors such as space availability, infrastructure requirements, and regulatory considerations may influence capital costs. While EC systems may entail higher initial capital investment than some conventional treatment technologies, such as chemical coagulation or sedimentation, they offer potential long-term cost savings through reduced chemical usage, lower energy consumption, and minimized sludge production. The operational costs are another significant component of the economic assessment for EC. These costs include expenses related to electricity consumption, electrode maintenance and replacement, chemical additives (if used), labor, and waste disposal [22]. Electricity consumption is a major operational cost for EC systems, as energy is required to power the electrodes and facilitate the electrochemical reactions. Strategies for optimizing energy consumption and reducing operational costs may include selecting energy-efficient equipment, optimizing process parameters, implementing control strategies, and exploring opportunities for renewable energy integration. Additionally, ongoing maintenance and periodic replacement of electrodes are necessary to ensure the efficient operation of EC systems and prevent performance degradation over time. Proper electrode maintenance practices, such as cleaning, inspection, and electrode surface renewal, can help extend their service life and minimize replacement costs. Cost-effectiveness analysis is an essential component of economic assessment for EC, and it is performed by comparing the total costs of implementing EC systems to the benefits achieved in terms of pollutant removal, water quality improvement, regulatory compliance, and other relevant factors. Cost-effectiveness metrics such as cost per unit volume of treated water, cost per unit mass of pollutant removed, or cost per unit of pollutant reduced can provide valuable insights into the economic efficiency of EC compared to alternative treatment technologies. By evaluating the costs and benefits of EC systems over their operational lifespan, stakeholders can determine whether EC represents a financially viable and competitive option for wastewater treatment. Return on investment (ROI) analysis is another important aspect of economic assessment for EC, estimating the financial returns generated by investing in EC systems relative to the initial capital outlay. ROI calculations typically consider factors such as operating revenues, cost savings, avoided penalties or fines, and other economic benefits associated with EC. The payback period, or the time required for the cumulative financial returns to equal the initial investment, is often used as a measure of investment profitability and risk. Shorter payback periods indicate higher ROI and greater financial attractiveness of EC investments. By conducting an ROI analysis, stakeholders can assess the economic viability of investing in EC systems and make informed decisions regarding resource allocation and project prioritization. Furthermore, the economic assessment of EC should consider the total cost of ownership (TCO), encompassing all costs incurred throughout the life cycle of EC systems, including capital costs, operational costs, maintenance costs, and disposal costs. TCO analysis provides a comprehensive view of the economic implications of EC investments, accounting for both direct and indirect expenses over time. By evaluating TCO alongside other economic metrics such as ROI and cost-effectiveness, stakeholders can gain a more holistic understanding of the financial implications of EC and identify opportunities for cost optimization and performance improvement [23].

9.3.1 Semi-Batch Process

The semi-batch EC process represents a hybrid approach to wastewater treatment that combines the benefits of both batch and continuous EC systems. In semi-batch EC, wastewater is treated in discrete batches, allowing for greater control over the treatment process and enhanced flexibility in operation compared to continuous systems. This paragraph explores the key principles, advantages, challenges, and applications of the semi-batch EC process, highlighting its potential as a versatile and efficient solution for water pollution control. The semi-batch EC process involves periodically introducing wastewater into a treatment reactor equipped with electrodes, where electrochemical reactions induce coagulation and precipitation of contaminants [24]. Unlike continuous EC systems, which operate continuously with a constant flow of wastewater, semi-batch systems treat wastewater in batches or cycles, allowing for intermittent operation and sequential treatment of discrete volumes of wastewater. During each treatment cycle, wastewater is introduced into the reactor and EC is initiated by applying an electrical current to the electrodes. As coagulation occurs, contaminants are removed from the wastewater through precipitation, flocculation, and settling processes. Once the treatment cycle is complete, the treated effluent is discharged and any residual sludge or by-products are removed from the reactor to prepare for the next batch. The primary advantages of the semi-batch EC process are its flexibility and adaptability to varying wastewater characteristics, flow rates, and treatment objectives. Unlike continuous systems, which require constant inflow and outflow of wastewater, semi-batch systems allow for intermittent operation, enabling operators to adjust treatment parameters, electrode configurations, and process conditions to meet changing treatment requirements. Semi-batch EC offers greater control over treatment kinetics, residence times, and mixing conditions than traditional batch or continuous systems, leading to improved treatment efficiency and performance. Furthermore, a semi-batch operation can facilitate the treatment of highly variable wastewater streams, intermittent pollutant sources, or batch-wise production processes, making it well-suited for diverse industrial, municipal, and environmental applications. Furthermore, semi-batch EC offers advantages in terms of process optimization, energy efficiency, and resource utilization. By treating wastewater in discrete batches, semi-batch systems allow for targeted dosing of chemical additives, optimization of electrode configurations, and minimization of energy consumption compared to continuous operation [25]. Operators can adjust treatment parameters such as current density, electrode spacing, reaction time, and pH to optimize coagulation efficiency and maximize contaminant removal while minimizing operational costs. Semi-batch EC systems can also be integrated with advanced process control strategies, real-time monitoring, and automation technologies to enhance operational efficiency, reliability, and performance. By leveraging these advantages, semi-batch EC can offer a cost-effective and sustainable solution for water pollution control and resource recovery. However, the semi-batch EC process also presents challenges and considerations that must be addressed to ensure effective and reliable operation. These challenges may be related to reactor design, electrode maintenance, sludge handling, and by-product management, which require careful planning and operational expertise to address them effectively. Moreover, the integration of semi-batch EC with other treatment processes and technologies presents opportunities for enhancing treatment efficiency, expanding treatment capabilities, and achieving synergistic benefits. In conclusion, the semi-batch EC process offers a versatile and efficient approach to wastewater treatment that combines the benefits of both batch and continuous systems. By treating wastewater in discrete batches, semi-batch EC allows for greater control over treatment parameters, enhanced flexibility in operation, and improved treatment efficiency compared to traditional batch or continuous systems [26].

9.3.2 Continuous Process

The continuous EC process represents an innovative approach to wastewater treatment that offers several advantages over traditional batch processes. In continuous EC, wastewater flows continuously through a treatment reactor equipped with electrodes, where electrochemical reactions induce coagulation and precipitation of contaminants. The continuous EC process operates on the principle of electrochemical coagulation, where an electrical current is applied to electrodes immersed in wastewater to generate coagulant species that destabilize and remove contaminants. In continuous EC systems, wastewater enters the treatment reactor at one end and flows continuously through the reactor, passing between the electrodes, where electrochemical reactions occur. As the wastewater comes into contact with the electrodes, metal ions are released into the solution, forming metal hydroxide or metal oxide coagulants that neutralize charged particles and facilitate their aggregation and precipitation. The treated effluent then exits the reactor while removing any residual sludge or by-products from the system to maintain treatment efficiency [27].

The primary advantage of continuous EC is its ability to provide consistent, reliable treatment over extended periods compared to batch processes. Continuous operation allows for steady-state conditions, uniform treatment kinetics, and consistent removal efficiencies, resulting in stable treatment performance and effluent quality. Unlike batch processes, which require intermittent loading, operation, and unloading of treatment reactors, continuous EC systems can treat wastewater continuously, making them well-suited for high-volume applications, variable flow rates, and continuous pollutant sources. Additionally, continuous EC offers advantages in terms of process automation, control, and monitoring, enabling real-time adjustment of treatment parameters to optimize performance and efficiency. Furthermore, continuous EC offers benefits in terms of energy efficiency, resource utilization, and operational cost savings compared to batch processes. Continuous systems can be designed with optimized electrode configurations, reactor geometries, and hydraulic conditions to minimize energy consumption while maximizing treatment efficacy. However, the continuous EC process also presents challenges and considerations that must be addressed to ensure effective and reliable operation [28]. One of the challenges is the electrode fouling, scaling, or degradation over time, which can reduce treatment efficiency and increase maintenance requirements. Electrode fouling may result from the accumulation of precipitated solids, organic matter, or corrosion products on electrode surfaces, leading to decreased electrode performance and increased energy consumption. In addition, continuous EC systems may require periodic cleaning, maintenance, and replacement of electrodes to maintain optimal treatment performance and prolong equipment lifespan. By treating wastewater continuously, continuous EC systems can provide stable treatment performance, optimize energy efficiency, and minimize operational costs while meeting stringent treatment requirements [29].

9.4 FEASIBILITY ANALYSIS OF SCALING UP THE EC PROCESS

Feasibility analysis of scaling up the EC process is essential for evaluating the technical, economic, and operational aspects of implementing larger-scale EC systems for wastewater treatment. As EC gains traction as a sustainable and effective water treatment technology, stakeholders need to assess the feasibility of scaling up EC processes to meet the demands of various industrial, municipal, and environmental applications. This chapter explores the key factors influencing the feasibility of scaling up EC, including technical considerations, economic implications, regulatory requirements, and operational challenges. The primary technical considerations in scaling up EC processes are system design and engineering. As EC systems increase in scale, factors such as reactor geometry, electrode configuration, electrode material, power supply capacity, and hydraulic conditions become more critical. Designing large-scale EC systems requires careful consideration of these factors to ensure optimal performance, efficiency, and reliability [3]. Additionally, scaling up EC processes may involve challenges related to mass transfer, electrode spacing, current distribution, and reactor hydrodynamics, which can impact treatment efficacy and energy efficiency. Computational modeling, pilot-scale testing, and experimental studies are valuable tools for evaluating the technical feasibility of scaling up EC processes and optimizing system design parameters to achieve desired treatment outcomes. Furthermore, economic feasibility analysis is essential for assessing the cost implications of scaling up EC processes. While larger-scale EC systems may offer economies of scale and potential cost savings compared to smaller installations, they also entail higher initial capital investment, operational costs, and maintenance expenses. Moreover, regulatory requirements and compliance considerations play a crucial role in the feasibility analysis of scaling up EC processes [30].

Wastewater treatment facilities must comply with regulatory standards and discharge limits set by local, state, and federal authorities to protect human health and the environment. Scaling up EC systems may involve obtaining permits, approvals, or certifications, conducting environmental impact assessments, and demonstrating compliance with water quality criteria and effluent standards. Regulatory compliance encompasses various aspects such as pollutant removal efficiency, discharge limits, sludge management, monitoring requirements, and reporting obligations [31]. Additionally, operational challenges and implementation barriers may impact the feasibility of scaling up EC processes. These challenges may include system complexity, technology maturity, reliability, operator training, supply chain constraints, and market acceptance. Overcoming operational challenges requires collaboration between technology developers, equipment manufacturers, service providers, and end-users to address technical limitations, optimize process performance, and streamline deployment procedures. Pilot-scale demonstrations, field trials, and case studies can provide valuable insights into real-world implementation challenges and inform decision-making regarding the feasibility of scaling up EC processes for practical applications [32].

9.5 CHALLENGES AND FUTURE TRENDS

Challenges and future trends for the EC (EC) process encompass a broad spectrum of technical, economic, regulatory, and operational considerations that shape the evolution and adoption of this innovative wastewater treatment technology. While EC offers several advantages, including the effective removal of various contaminants and reduced reliance on chemical additives, it also faces challenges related to scalability, energy efficiency, by-product management, and regulatory compliance. Additionally, emerging trends such as technological advancements, process optimization, integration with other treatment methods, and market demand for sustainable solutions are shaping the future trajectory of EC. This paragraph explores the key challenges and future trends for EC, highlighting opportunities for innovation, improvement, and broader adoption. One of the primary challenges for EC is scalability, particularly in transitioning from lab-scale or pilot-scale demonstrations to larger industrial or municipal applications. Scaling up EC systems involves addressing technical issues such as reactor design, electrode configuration, hydraulic conditions, and energy consumption. Furthermore, energy efficiency is a key consideration for EC, as electricity consumption represents a significant operational cost and environmental impact. Moreover, by-product management presents a challenge for EC, as the treatment process generates secondary pollutants or residues that require proper handling and disposal. Common by-products of EC include metal hydroxide precipitates, hydrogen gas, residual dissolved ions, and electrode degradation products. Managing these by-products effectively requires strategies for sludge dewatering, stabilization, and disposal as well as addressing potential environmental and health risks associated with residual contaminants. Looking ahead, several future trends are shaping the evolution of EC and driving innovation in the field. Technological advancements, such as novel electrode materials, reactor designs, and control systems, are enhancing the performance, efficiency, and reliability of EC systems. Process optimization strategies, such as real-time monitoring, feedback control, and predictive modeling, improve treatment efficacy, energy efficiency, and operational flexibility. Integration of EC with other treatment methods, such as membrane filtration, advanced oxidation processes, or biological treatment, enhances the overall treatment efficiency and expands the range of contaminants that can be effectively removed. Additionally, market trends such as increasing demand for water reuse, resource recovery, and sustainable solutions are driving innovation in EC technology to meet evolving customer needs and market demands. In conclusion, challenges and future trends for EC encompass a diverse range of technical, economic, regulatory, and operational considerations that shape the development, adoption, and commercialization of this innovative wastewater treatment technology.

9.6 CONCLUSION

EC emerges as a highly promising technology for achieving both environmental and economic sustainability in wastewater treatment. By harnessing the power of electricity to remove contaminants from water sources, EC offers effective solutions to mitigate water pollution, protect ecosystems, and safeguard public health. Moreover, its operational efficiency, scalability, and potential for resource recovery make it a cost-effective alternative to conventional treatment methods, thereby promoting economic growth and resource conservation. As societies strive to address the challenges of water scarcity, pollution, and climate change, EC stands as a symbol of hope, offering a pathway toward a more sustainable and resilient future. Furthermore, the environmental assessment of the EC process is essential for understanding its potential impacts and benefits in wastewater treatment applications. By evaluating water quality improvements, energy consumption, greenhouse gas emissions, sludge generation, and the overall sustainability, stakeholders can make informed decisions regarding the adoption and optimization of EC technologies. Through continuous innovation, research, and collaboration, EC has the potential to play a significant role in advancing sustainable water management practices and addressing environmental challenges in the years to come.

REFERENCES

[1] S. Hube, M. Eskafi, K.F. Hrafnkelsdóttir, B. Bjarnadóttir, M.Á. Bjarnadóttir, S. Axelsdóttir, B. Wu, Direct membrane filtration for wastewater treatment and resource recovery: A review, *Sci. Total Environ.* 710 (2020). https://doi.org/10.1016/j.scitotenv.2019.136375.

[2] M. Bharti, P.P. Das, M.K. Purkait, A review on the treatment of water and wastewater by electrocoagulation process: Advances and emerging applications, *J. Environ. Chem. Eng.* 11 (2023) 111558. https://doi.org/10.1016/j.jece.2023.111558.
[3] P.P. Das, M. Sharma, M.K. Purkait, Recent progress on electrocoagulation process for wastewater treatment: A review, *Sep. Purif. Technol.* 292 (2022) 121058. https://doi.org/10.1016/j.seppur.2022.121058.
[4] J. Clavreul, H. Baumeister, T.H. Christensen, A. Damgaard, An environmental assessment system for environmental technologies, *Environ. Model. Softw.* 60 (2014) 18–30. https://doi.org/10.1016/j.envsoft.2014.06.007.
[5] S. Sala, F. Biganzoli, E.S. Mengual, E. Saouter, Toxicity impacts in the environmental footprint method: Calculation principles, *Int. J. Life Cycle Assess.* 27 (2022) 587–602. https://doi.org/10.1007/s11367-022-02033-0.
[6] T.A. Aragaw, Recycling electro-coagulated sludge from textile wastewater treatment plants as an adsorbent for the adsorptions of fluoride in an aqueous solution, *Heliyon* 7 (2021) e07281. https://doi.org/10.1016/j.heliyon.2021.e07281.
[7] G. Chen, P.L. Yue, A.S. Mujumdar, Sludge dewatering and drying, *Dry. Technol.* 20 (2002) 883–916. https://doi.org/10.1081/DRT-120003768.
[8] R.J. Wakeman, Separation technologies for sludge dewatering, *J. Hazard. Mater.* 144 (2007) 614–619. https://doi.org/10.1016/j.jhazmat.2007.01.084.
[9] V.H.P. To, T.V. Nguyen, S. Vigneswaran, H.H. Ngo, A review on sludge dewatering indices, *Water Sci. Technol.* 74 (2016) 1–16. https://doi.org/10.2166/wst.2016.102.
[10] J. Olivier, J. Vaxelaire, Municipal sludge dewatering by belt filter press: Effect of operating parameters, *J. Chem. Technol. Biotechnol.* 80 (2005) 948–953. https://doi.org/10.1002/jctb.1267.
[11] E.G. Hertwich, G.P. Peters, Carbon footprint of nations: A global, trade-linked analysis, *Environ. Sci. Technol.* 43 (2009) 6414–6420. https://doi.org/10.1021/es803496a.
[12] S.S. Muthu, Y. Li, J.Y. Hu, P.Y. Mok, Carbon footprint of shopping (grocery) bags in China, Hong Kong and India, *Atmos. Environ.* 45 (2011) 469–475. https://doi.org/10.1016/j.atmosenv.2010.09.054.
[13] S.S.K. Muthu, Y. Li, J.Y. Hu, L. Ze, Carbon footprint reduction in the textile process chain: Recycling of textile materials, *Fibers Polym.* 13 (2012) 1065–1070. https://doi.org/10.1007/s12221-012-1065-0.
[14] A.L. Andrady (Ed.), *Plastics and the Environment: A Handbook*, John Wiley & Sons, New York, 2003: pp. 697–725, pp. 1–27.
[15] S. Borzooei, G. Campo, A. Cerutti, L. Meucci, D. Panepinto, M. Ravina, V. Riggio, B. Ruffino, G. Scibilia, M. Zanetti, Feasibility analysis for reduction of carbon footprint in a wastewater treatment plant, *J. Clean. Prod.* 271 (2020). https://doi.org/10.1016/j.jclepro.2020.122526.
[16] S. Dhara, P.P. Das, R. Uppaluri, M.K. Purkait, Chapter 12—Biological approach for energy self-sufficiency of municipal wastewater treatment plants, in: M. Sillanpää, A. Khadir, K. Gurung (Eds.), *Resource Recovery in Municipal Waste Waters*, Elsevier, 2023: pp. 235–260. https://doi.org/10.1016/B978-0-323-99348-7.00006-0.
[17] S. Barbhuiya, B.B. Das, Life cycle assessment of construction materials: Methodologies, applications and future directions for sustainable decision-making, *Case Stud. Constr. Mater.* 19 (2023) e02326. https://doi.org/10.1016/j.cscm.2023.e02326.
[18] P.P. Das, A.D. Sontakke, M.K. Purkait, Chapter 2—Electrocoagulation process for wastewater treatment: Applications, challenges, and prospects, in: M.P. Shah (Ed.), *Development in Wastewater Treatment Research and Processes*, Elsevier, 2023: pp. 23–48. https://doi.org/10.1016/B978-0-323-95684-0.00015-4.
[19] D. Das, B.K. Nandi, Removal of Fe (II) ions from drinking water using Electrocoagulation (EC) process: Parametric optimization and kinetic study, *J. Environ. Chem. Eng.* 7 (2019) 103116. https://doi.org/10.1016/j.jece.2019.103116.

[20] D. Ghosh, H. Solanki, M.K. Purkait, Removal of Fe(II) from tap water by electrocoagulation technique, *J. Hazard. Mater.* 155 (2008) 135–143. https://doi.org/10.1016/j.jhazmat.2007.11.042.
[21] M. Bharti, P.P. Das, M.K. Purkait, A review on the treatment of water and wastewater by electrocoagulation process: Advances and emerging applications, *J. Environ. Chem. Eng.* 11 (2023). https://doi.org/10.1016/j.jece.2023.111558.
[22] R. Niazmand, M. Jahani, S. Kalantarian, Treatment of olive processing wastewater by electrocoagulation: An effectiveness and economic assessment, *J. Environ. Manage.* 248 (2019) 109262. https://doi.org/10.1016/j.jenvman.2019.109262.
[23] R. Mahmud, S.M. Moni, K. High, M. Carbajales-Dale, Integration of techno-economic analysis and life cycle assessment for sustainable process design—a review, *J. Clean. Prod.* 317 (2021) 128247. https://doi.org/10.1016/j.jclepro.2021.128247.
[24] K. Rajaniemi, M. Raulio, S. Tuomikoski, U. Lassi, Comparison of batch and novel continuous electrocoagulation processes in the treatment of paint industry wash water, *Desalin. Water Treat.* 170 (2019) 394–404. https://doi.org/10.5004/dwt.2019.24562.
[25] T.T.Q. Nguyen, P. Loganathan, B.K. Dinh, T.V. Nguyen, S. Vigneswaran, H.H. Ngo, Removing arsenate from water using batch and continuous-flow electrocoagulation with diverse power sources, *J. Water Process Eng.* 41 (2021) 102028. https://doi.org/10.1016/j.jwpe.2021.102028.
[26] A.R. Vargas, C.S. Guillen, M.E. Magaña Haynes, F.Y. AlJaberi, Nickel removal from an industrial effluent by electrocoagulation in semi-continuous operation: Hydrodynamic, kinetic and cost analysis, *Results Eng.* 17 (2023) 100961. https://doi.org/10.1016/j.rineng.2023.100961.
[27] M. López-Guzmán, M.A. Flores-Hidalgo, L. Reynoso-Cuevas, Electrocoagulation process: An approach to continuous processes, reactors design, pharmaceuticals removal, and hybrid systems—a review, *Processes* 9 (2021). https://doi.org/10.3390/pr9101831.
[28] M. Kobya, E. Gengec, E. Demirbas, Operating parameters and costs assessments of a real dyehouse wastewater effluent treated by a continuous electrocoagulation process, *Chem. Eng. Process. Process Intensif.* 101 (2016) 87–100. https://doi.org/10.1016/j.cep.2015.11.012.
[29] A.R. Makwana, M.M. Ahammed, Continuous electrocoagulation process for the post-treatment of anaerobically treated municipal wastewater, *Process Saf. Environ. Prot.* 102 (2016) 724–733. https://doi.org/10.1016/j.psep.2016.06.005.
[30] A. Omoregie, *A Feasibility Study to Scale-Up the Production of Sporosarcina pasteurii, Using Industrial-Grade Reagents, for Cost-Effective in-Situ Biocementation,* Thesis, Swinburne University of Technology, 2020.
[31] O. Pahl, A. Firth, I. MacLeod, J. Baird, Anaerobic co-digestion of mechanically biologically treated municipal waste with primary sewage sludge—a feasibility study, *Bioresour. Technol.* 99 (2008) 3354–3364. https://doi.org/10.1016/j.biortech.2007.08.027.
[32] V. Kuokkanen, T. Kuokkanen, J. Rämö, U. Lassi, J. Roininen, Removal of phosphate from wastewaters for further utilization using electrocoagulation with hybrid electrodes—techno-economic studies, *J. Water Process Eng.* 8 (2015) e50–e57. https://doi.org/10.1016/j.jwpe.2014.11.008.

10 Advancement and Future Research Scope

10.1 INTRODUCTION

People have been suffering from a lack of water and declining water quality for a very long time. One of the main focuses of the Sustainable Development Goals (SDGs) of the United Nations is now access to clean drinking water. It is estimated that 1.2 billion people do not have access to clean drinking water. Additionally, a growing amount of toxins from human activity are entering water due to the extensive usage of water. Traditional pollutants like colloids, nutrients, microorganisms, and heavy metals are among them. Emerging pollutants like pharmaceuticals and perfluorinated compounds (PFCs) are also included in this category. Among these, a lot of pollutants are typically challenging to handle using traditional treatment methods. Thus, there is a pressing need for affordable, environmentally friendly, and sustainable water treatment methods [1].

In recent years, electrocoagulation (EC), also known as electroflocculation, has gained recognition as a viable and effective substitute method for water treatment. It is interesting to note that EC has many benefits, such as easy automation, decreased sludge formation, and straightforward operation [2–4]. Sacrificial anodes are used in the EC process to produce coagulants *in situ*, which is an effective way to remediate a variety of contaminants in water and wastewater. The primary mechanisms of the EC process for removing pollutants include adsorption, charge neutralization, and entanglement in precipitation (sweep coagulation) [5]. Many researchers have been drawn to this field of study because, in addition to its effectiveness and capacity to eliminate a variety of contaminants, it also has the advantage of having a low operating cost, being easy to use, and having the capacity to neutralize the pH of the solution during electrolysis. The primary disadvantage of EC is its need for electrical energy to power the process, especially when dealing with wastewater with poor conductivity [6–8]. Over the past few decades, there has been an increase in the study of EC processes due to the process's advantages and the growing interest in electrified processes. Thus, scientists have evaluated how well EC performs in treating industrial wastewater, especially pharmaceutical wastewater, distillery effluents, and other developing toxins including arsenic, fluoride, and pesticides. Reviews of research on reactor modeling, EC reactor design, passivation and its controlling mechanisms, and the application of EC in conjunction with other water and wastewater treatment procedures have also been conducted. EC has proved to be successful in cleaning up a variety of contaminants; however, there are still certain restrictions. For instance, it is necessary to take into account the generation of undesirable by-products, the remediation of wastewater including mixed pollutants, and the dilution of contaminants

DOI: 10.1201/9781003499671-10

(micropollutants) [9]. It has been demonstrated that EC works well to remove inorganic pollutants from produced water used in the oil and gas sector, including metal cations, silica, and hardness (i.e., Ca and Mg). Nevertheless, it was discovered that the amount of organic pollutants removed from these effluents after EC treatment was restricted. After EC, additional post-treatment is frequently needed to eliminate certain contaminants. Moreover, in some circumstances the use of EC may be limited by the cost of electrical energy usage. Therefore, it is preferable to improve EC to treat a larger range of pollutants or use less energy [10, 11].

This chapter reports on a number of EC advancements, such as the use of sustainable power sources, self-powered air cathode EC, and the aerated EC process. Additionally, a full discussion of the impact of H_2O_2 and the kind of electric current (alternating and direct) on the EC process has been held. Additionally, a detailed discussion of the creation of hydrogen gas, the control of electrocoagulated sludge, and the design of electrochemical reactors using a CFD approach has been provided. In summary, this chapter concentrates on these developments and the scope of the future EC process study.

10.2 NEW INSIGHTS ON THE ELECTROCOAGULATION PROCESS

10.2.1 Aerated Electrocoagulation System

Aeration has improved performance to the point where it is now a standard operating parameter in EC treatment. Because of the depletion of dissolved oxygen, the removal effectiveness of EC using iron electrodes diminishes with time in the absence of aeration. This demonstrates how the dissolved oxygen consumption has an impact on the EC process's removal mechanisms and overall performance [12–14]. The iron EC's efficacy was assessed by Syam Babu et al. (2021) both with and without external aeration [15]. A synthetic arsenite solution was treated utilizing the procedure in a rectangular 5 L batch reactor equipped with 14 iron electrodes. The authors noted that external aeration greatly enhanced the EC efficacy for eliminating arsenic from an aqueous solution. Ferrell ions and other oxidants may be produced by the aqueous solution's increasing dissolved oxygen content, which is a result of external aeration. Thus, the *in situ* synthesis of oxidizing agents like ferro ions in the solution has the advantage of improving the process for removing arsenic from the solution. In another study, Syam Babu et al. (2021) showed how to use iron EC and continuous aeration to remove arsenite from simulated arsenite-containing groundwater [16]. One mg L^{-1} of arsenite was treated for up to 12 six-hour cycles at 10 and 5 L h^{-1}, as well as with and without altering polarity. The first four cycles at a flow rate of 10 L h^{-1} resulted in the almost total removal of arsenite. Consequently, there was less flow of ferrous ions into the solution from the anode surface.

According to van Genuchten et al. (2016), low dissolved oxygen in iron EC causes iron electrodes to passivate quickly, which lowers treatment efficacy [17]. Furthermore, increasing the amount of air in the system causes more contaminants to collide with the floc, which improves the removal of pollutants. Conversely, at the cathode, H^+ ion consumption and oxygen reduction take place. As a result, external aeration consumes more hydrogen ions than a non-aerated solution does, which raises the pH.

As a result of the higher pH established in the solution and the enhanced aeration action, stronger coagulants are generated. Accordingly, recent research has shown that external aeration during iron EC treatment greatly improves process efficiency and lowers passivation.

10.2.2 Air Cathode Electrocoagulation Process

The cost and environmental effects of the electrical energy necessary for EC remediation of wastewater treatment, especially for low conductivity streams, provide a significant problem. While adding external aeration to EC increases performance, it also consumes more energy. To increase the amount of dissolved oxygen available inside the EC system, high aeration rates are needed. A self-powered EC system is now possible owing to the air cathode EC process (ACEC), which replaced the metal cathode in EC with an air diffusion cathode. Rather than hydrogen development from reduction of water, the process at the air cathode is the oxygen reduction reaction (ORR). Because oxidizing species are produced at the air cathode, this designed technique improves pollutant removal and reduces energy costs associated with EC operation. To effectively remove impurities while generating power, ACEC combines electrochemical advanced oxidation and other processes [18–20].

$$O_2 + 2e^- + 2H^+ \rightarrow H_2O_2 E^0 = 0.683V \quad (10.1)$$

$$O_2 + 4e^- + 4H^+ \rightarrow 2H_2O E^0 = 0.816V \quad (10.2)$$

$$Fe(s) \rightarrow Fe^{2+} + 2e^- E^0 = -0.447V \quad (10.3)$$

The electrode material is one of the many variables influencing the performance of the ACEC process; it controls the efficiency of pollution removal and energy production. For instance, the ORR catalyst affects the cathode's capacity to produce H_2O_2 for the Fenton reaction, which creates hydroxyl radicals. The anode material affects the floc's ability to adsorb during the EC process [21].

$$Fe^{2+} + H_2O_2 \rightarrow \bullet OH + Fe^{3+} + OH^- \quad (10.4)$$

Xu et al. (2020) discovered that the cathode materials made of iron, nickel, and carbon fiber had varying energy production and removal efficiencies [22]. However, because carbon fiber has a strong oxygen reduction performance, more energy was produced utilizing it. In ACEC, the trade-off between energy consumption and pollution removal results in synergistic benefits. The primary reason why ACEC is better at removing pollutants than other EC systems is because it produces more H_2O_2, which in turn produces potent oxidants. The process is also impacted by the makeup of the water that needs to be treated. It has been assessed whether various salts such as NaCl, Na_2SO_4, and $NaHCO_3$ are present. Because chloride ions produce active chlorine and prevent oxide layer formation on the anode, it has been discovered that using them in ACEC results in superior contaminant removal. Compared to $NaHCO_3$, which was shown to restrict energy generation due to the creation of a passive layer on the anode, NaCl and Na_2SO_4 enabled increased energy production. Furthermore, because of the reduction in overpotential and ohmic loss at the electrode surface, energy output was proportionate to the concentration of the electrolyte. It was also discovered that the initial pH had a significant impact on floc development and energy output. For ACEC, a neutral pH in the range of 4–5 is ideal since it promotes the development of more flocs [23, 24]. For phosphor recovery, Kim et al. (2018) employed a magnesium anode [25]. There was a high generation of energy. Due to the magnesium anode's easy electron loss, a comparatively significant amount of energy is produced. It has been discovered that ACEC effectively removes pollutants while generating energy, resolving the issue of EC's electrical energy consumption. There are still a few obstacles in the way of using this technology, though. Polarity reversal is impractical when using an air cathode; alternative strategies for reducing anode fouling may therefore need to be used in ACEC procedures, which is contingent upon the nature and amount of water being treated.

10.2.3 Effect of H_2O_2 on Electrocoagulation

High pollutant removal performance in water remediation has been achieved by combining EC technology with peroxide-assisted treatment (ECPAT). Numerous studies that have examined the benefits and drawbacks of EC-aided advanced oxidation have come to the conclusion that the addition of H_2O_2 enhances the therapeutic process. This is particularly true for organic cleanup, as the process encourages the production of extremely oxidizing hydroxyl radicals (•OH). The procedure is inexpensive, easy to use, and has little risks. Ozone is dangerous as well, necessitating the use of an ozone destruction machine. When organic contaminants are present in water that is being treated, the hydroxyl radicals have the ability to oxidize and break down their structural integrity. Additionally, the existence of iron species produced in EC by H_2O_2 facilitates Fenton's process, which is a series of oxidizing processes that ferrous (II) ions catalyze [26, 27]. There are two possible mechanisms for this reaction: either the generation of highly oxidizing Fe(IV) species called ferryl ions (FeO^{2+}) or the free radical chain reactions proposed by Haber-Weiss [28]. H_2O_2 was also found to affect the efficiency of aluminum's pollutant removal process, despite the metal having only one oxidation state (+3). According to Roa-Morales et al. (2007), the addition of H_2O_2 enhanced COD removal using Al-EC at pH 4 from 80% to 90% [29]. This boost is most likely the result of chain events that begin when H_2O_2 is reduced cathodically to produce hydroxyl radicals (Eq. 10.5).

$$H_2O_2 + e^- + H^+ \rightarrow \bullet OH + H_2O \tag{10.5}$$

Finding the ideal H_2O_2 dose for an EC-PAT process is crucial since too much H_2O_2 might cause radical scavenging and lower efficiency. The effectiveness of EC-PAT may also be impacted by pH. For instance, pH> 8 is suggested for effective EC performance based on floc production with iron electrodes, whereas pH< 3 is highly active for Fenton's process. As a result, it could be advantageous to adjust the pH level throughout the procedure, eliminating organic pollutants at lower pH levels via oxidation and then forming floc at higher pH levels. Although the addition of peroxide has been demonstrated to improve the removal of organic pollutants during EC, it is unclear whether this effect is synergistic—that is, whether the same result might be obtained using the two processes in succession [10].

10.2.4 Sinusoidal Alternating Current Electrocoagulation (AC-EC)

EC has been operated with sinusoidal alternating current (AC-EC) in a number of studies in an effort to increase efficiency and decrease passivation. Every electrode in an AC-EC system has the potential to be an interchangeable anode or cathode since the anode and cathode are alternated on a periodic basis according to the sinusoidal waveform. The majority of the time, AC-EC has been tested at 50 Hz using diverse electrode materials for a range of purposes. According to the majority of writers, AC-EC and DC-EC remove pollutants similarly, with AC-EC having a tiny edge. The periodic switching of the anode and cathode activities, which lessens the strain on each electrode's functionality, delays passivation, and enhances metal (oxy)hydroxide production, may be the cause of higher removal efficiencies in AC-EC [30]. Furthermore, like APC-EC, AC-EC's metallic cations disperse more evenly throughout the cell as a result of anode and cathode corrosion, which enhances efficiency by facilitating coagulant mass transfer and coagulant–pollutant interactions.

Different electrode materials passivate differently due to the dependence of the electrode material's mechanism and rate of passivation. Because of this, depending on the electrode material, using an AC waveform to avoid electrode passivation may provide varying outcomes. Additionally, the literature suggests that the distinctions between APC-EC and DC-EC are more significant than the advances in pollutant removal efficiency in AC-EC relative to DC-EC [31, 32]. Under some experimental conditions, the PR time in AC-EC may be less than the time needed to charge the electrode double layer, rendering this technique ineffective, particularly when working with fluids that have low conductivity. Another factor leading to the lower removal rate of contaminants in AC-EC compared to DC-EC could be lower faradaic efficiency. The majority of research suggests that ACEC is more efficient than DC-EC in terms of energy usage. Compared to DC-EC, AC-EC used between 13 and 80% less energy. In AC-EC, however, switching the polarity of the anode and the cathode lessens the formation of the passivating coating. However, a few studies have found that although the differences were not statistically significant, AC-EC energy usage tended to be higher than DC-EC energy consumption [33]. For example, to remove nitrate from water, DC-EC performed marginally better in terms of energy efficiency than AC-EC at current intensities ranging from 2 to 4 A. The different types of pollutants and their complexities of removal procedures may be related to the increased energy usage in AC-EC. According to all the analyzed studies, electrode consumption is reduced when the AC waveform is utilized rather than the DC waveform. It has been demonstrated that, independent of the type of electrode material utilized, electrode consumption in AC-EC can be lowered by around 14–93% when compared to DC-EC [34]. It was demonstrated by scanning electron microscope (SEM) imaging that the unidirectional current causes electrode oxidation in DC-EC to occur at the same preferred areas of the electrode surface. As a result, the electrode corrodes unevenly and randomly. On the other hand, with AC-EC, the returning current creates new oxidation spots that increase the evenly attacked area while decreasing the locally attacked portion of the anode/cathode surface. As a result, compared to electrodes in AC-EC, those in DC-EC are more vulnerable to localized corrosion and mineral precipitation. Consequently, the electrode mass loss decreases as the type of corrosion shifts from pitting corrosion in DC-EC to uniform corrosion in AC-EC. Moreover, side reactions—which deflect current away from electrode corrosion—are more likely to occur in AC-EC when there are frequent changes in electrode polarity. Because the anode and the cathode dissolve uniformly and require less electrode consumption when utilizing the AC waveform rather than the DC waveform, EC experiments have demonstrated a decrease in sludge generation. For instance, it has been observed that switching the current waveform from DC to AC might reduce sludge generation by as much as 54%. Consequently, AC-EC is more economical than DC-EC as the primary operational costs of EC are energy, electrode use, and sludge disposal [10, 35].

10.3 RENEWABLE ENERGY SOURCE FOR POWER SUPPLY IN ELECTROCOAGULATION

There is growing interest in using solar energy or microbial fuel cells (MFC) as an energy source for electrochemical processes (ECPs). García-Orozco et al. (2022) treated wastewater from the chocolate business using solar-photovoltaic EC [36]. Using aluminum electrodes for EC produced good results, removing BOD and COD with an efficiency of 39% and 50%, respectively. Furthermore, the energy savings were calculated and showed that the EC method might save a substantial amount of money. Similar to this, Mohamad et al. (2022) treated palm oil effluent by directly attaching a photovoltaic solar module to the EC system [37]. The treatment yielded 90% removal efficiency for BOD and COD in this integrated system. The initial concentrations of pollutants were 16 g L^{-1} BOD and 26 g L^{-1} COD. The concentrations were lowered to 2.1 g L^{-1} COD and 0.46 g L^{-1} BOD after 480 minutes of EC. The findings showed that the PV system's power generation and current intensity were mostly responsible for the pollutants' elimination. Therefore, it might be suggested that the EC process be directly powered by solar energy as a practical means of lowering the amount of external electrical energy consumed. Microbial fuel cells (MFCs) are a novel bioelectrochemical device that uses electrons from bacterially catalyzed metabolic reactions to produce power. An anaerobic anodic chamber and an aerobic cathodic chamber are the standard components of MFCs. Electrogenic bacteria generate bioelectricity through a redox process. The idea of employing MFCs to change an energy-intensive EC process into an energy-neutral process for treating water contaminated by algae was illustrated by Dong et al. (2017) [38]. To eliminate algal cells at various influent conductivities, a biocathode-coupled electrocoagulation cell (bio-EC) was developed, featuring a nitrifying biocathode and a sacrificial iron anode. The nitrogen removal rates of 7.28 mg L^{-1} h^{-1} and 6.77 mg L^{-1} h^{-1} were found to be associated with NH^{4+} and N removal efficiencies of 55.7 ± 2.1% and 51.7 ± 2.3%, respectively, as reported using the bio-EC. The findings imply that an essential part of the process's algal elimination was carried out using the nitrifying biocathode.

10.4 FUTURE RESEARCH

10.4.1 Reactor Design and Key Scale-Up Parameters

The arrangement and connection of the electrodes in an EC system have a significant impact on the system's overall performance. Particularly, electrode configuration—that is, its shape, pattern, etc.—has attracted a lot of attention from researchers recently and is a topic of general public concern. Series or parallel electrode connections can be used in either a monopolar or bipolar system. Electrode connection techniques affect process cost and energy usage in addition to removal efficiency. Monopolar electrodes in parallel connections (MP-P), monopolar electrodes in serial connections (MP-S), and bipolar electrodes in serial connections (BP-S) are the three most often used electrode configurations [39]. All of the anodes and cathodes in the MP-P mode are connected to one another in the same way. The interior electrode, sometimes known as the "sacrificial electrode," may be composed of the same metal or a different one. Lowering the anode's consumption and passivating the cathode are its primary functions. The global voltage in the MP-S mode is the sum of the voltages in each individual electrolytic cell, yet the electric current flowing through each electrode is the same. When using the BP-S mode, the external power supply is directly linked to the outermost electrode, while the inside electrodes are left unconnected. The internal electrode's surrounding sides become polarized and transfer opposing charges to the surrounding electrodes as soon as the current flows through the main electrode. As a result, the inside electrodes are bipolar, while the two outermost electrodes are unipolar [40]. This suggests that every internal electrode functions as both a cathode (the other side) and an anode (one side) at the same time. In actuality, electrode designs are versatile and varied in terms of shape and pattern. Novel electrodes such as perforated, cylindrical, spherical, netting, and revolving electrodes have been produced, building on the foundation of solid plate electrodes. These innovative electrodes are generally designed to decrease electrode passivation and, as a result, energy consumption and improve mass movement and mixing efficiency. E. coli was eliminated from wastewater by Hashim et al. (2020) using an EC reactor with perforated Al plates [41]. They discovered that using perforated electrode plates cut down on running expenses and the requirement for mechanical or magnetic stirrers. The removal of 96% of E. coli found in wastewater is possible under ideal working conditions. Khandegar and Saroha (2016) treated synthetic dye wastewater with EC system equipped with Al electrodes and analyzed how holes—including number, diameter, and shape—affected the effectiveness of color removal [42]. The color removal efficiency was found to be enhanced by EC with punched electrodes as opposed to plane electrodes. It was discovered that when the diameter and number of holes were increased, the EC performance increased up to an optimal value. In order to lessen cathode passivation, Ibrahim et al. (2020) treated synthetically created water using EC reactors equipped with perforated Fe and Al electrodes [43]. The findings demonstrated that EC with a cylindrical perforated electrode was superior to a standard plate electrode in terms of minimizing cathode passivation and reducing energy usage by almost 70%. Abdulhadi et al.'s (2021) design of an EC reactor with perforated electrodes is shown in Figure 10.1(a) [44]. An example of a turbulent flow reactor with perforated electrodes is shown in Figure 10.1(b). With current reversal, three plates function as cathodes and three plates as anodes at the same time. The reactor's turbulent flow is encouraged by punctured electrodes, which raises the EC efficiency. Perforated plates have superior mass transfer and treatment performance to solid plates. Nevertheless, they often require high pressure to function and can result in electrical short circuits. The economic analysis of EC reactors with perforated plates must be taken into systematic consideration when they are used in actual engineering. Figure 10.1(c)–(h) display several types of reactors with different electrode designs. A particular kind of cylindrical cathode reactor with a wire netting sheet anode is depicted in Figure 10.1(c). This reactor is composed of a ferroelectric cylindrical cathode. Its anode, which is situated in the reactor's center, is a sheet of cylindrically wound Fe hexagonal wire netting. A cylindrical cathode reactor with Al ball anodes is shown in Figure 10.1(d) [1]. The outer layer contains the cathode that has circular holes in it. A Plexiglass cylinder containing ingredients for the Al ball anode forms the inner layer. Ball anodes have some advantages over plate-shaped electrodes, including greater removal efficiency, convenience of usage, and the ability to accommodate more anodes with a higher surface area. Tezcan Un et al. (2013) designed a cylindrical anode reactor with a revolving impeller cathode (Figure 10.1(e)) [45]. The reactor's central component, the revolving impeller cathode, was made up of two Al blades. During EC, the

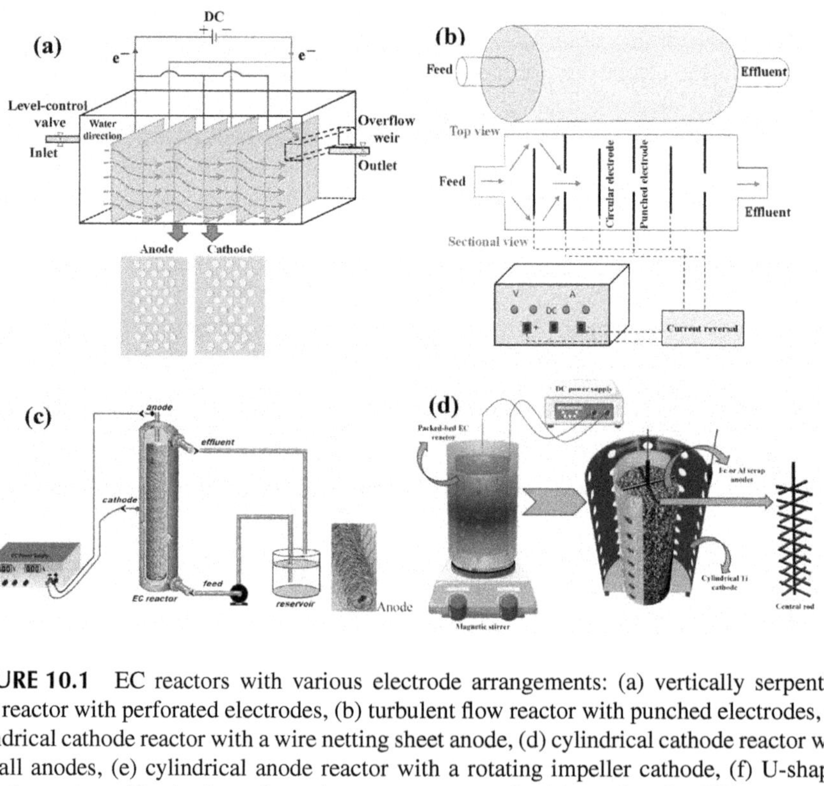

FIGURE 10.1 EC reactors with various electrode arrangements: (a) vertically serpentine flow reactor with perforated electrodes, (b) turbulent flow reactor with punched electrodes, (c) cylindrical cathode reactor with a wire netting sheet anode, (d) cylindrical cathode reactor with Al ball anodes, (e) cylindrical anode reactor with a rotating impeller cathode, (f) U-shaped cathode reactor with a horizontal rotating screw-type anode, (g) rotating disc EC reactor, and (h) filter-press flow EC reactor with monopolar connection (FM01-LC type) (Reproduced with permission from Lu et al. (2021) @ Elsevier [1]).

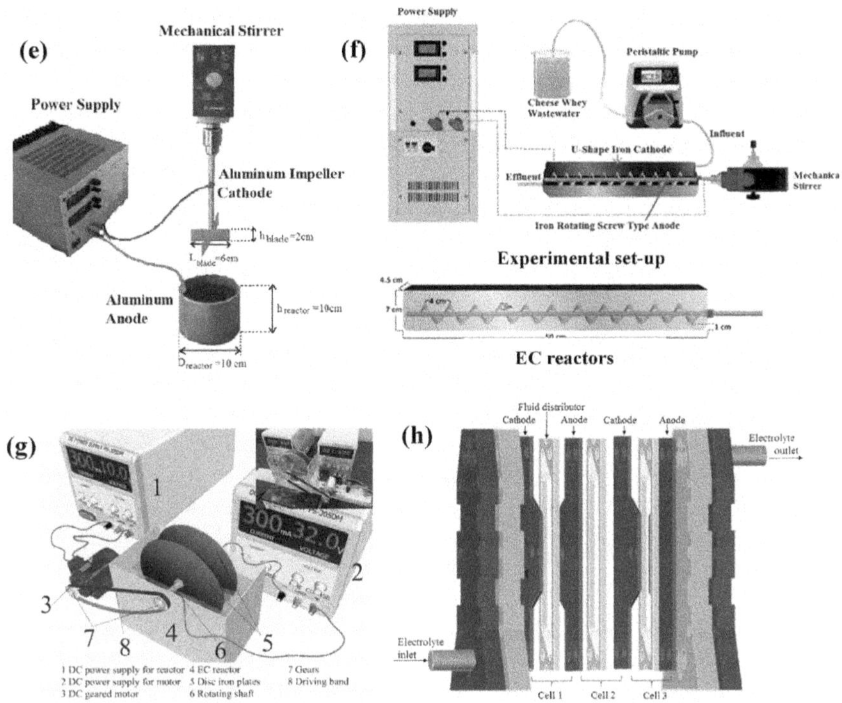

FIGURE 10.1 (Continued).

solution was mechanically agitated by the moving impeller to preserve homogeneity and stop the particles from settling. In a similar vein, they advanced the development of a continuous EC reactor including a U-shaped Fe cathode and a horizontally spinning screw-type Fe anode (Figure 10.1(f)). Because of the induced convection in the bulk, it was found that this reactor exhibited high mass transfer [46]. Xu et al. (2018) investigated the use of a rotating disc EC reactor to lower the passivation rate of electrodes (Figure 10.1(g)) [47]. In this reactor, two parallel Fe disc electrodes operating in monopolar mode were vertically arranged and placed on a revolving shaft, with a surface area that was briefly submerged. The findings showed that when the rotation speed increased, the passivation rate decreased, enhancing the EC performance. The filter-press cell is another popular type of electrochemical reactor used in EC. As seen in Figure 10.1(h), the FM01-LC cell is a type of filter-press reactor that is widely employed in laboratory-scale testing. Because filter-press reactors are simple to scale up, they can be used in both laboratories and factories.

Turbulent eddies, including jet flows, vortices, and rotating flow structures, are produced in filter-press flow reactors with multi-electrode stacks by random velocity variations caused by changes in the cross-sectional area of the water flow. As a result, filter-press flow reactors have a number of very significant benefits, including high mass transport, uniform current density distribution, and straightforward design [48].

In essence, the reported design is based on pilot-, laboratory-, or industrial-scale studies. The EC reactor's operation, performance, cost-effectiveness, and scale-up are all impacted by its rational design. One crucial component of reactor design is the reactor's shape. The majority of EC reactors have a rectangular or cylindrical top view, according to an analysis of their shapes. Water flow in an EC reactor is primarily classified into two types based on its orientation: vertical and horizontal [40]. Rectangular and cylindrical reactors have been commonly employed by researchers in batch experiments, as shown in Figure 10.2(a) and (b). Because the electrode plates are easily replaceable, vertical electrodes are more commonly employed in EC reactors than horizontal electrodes (Figure 10.2(c) and (d)). A common continuous flow reactor with perforated input pipes—which occasionally have both main and branch pipes—is shown in Figure 10.2(c). It is possible to improve influent uniformity and mixing efficiency as well as modify the flow pattern in reactors by arranging the plates differently by putting perforated inlet pipes in the EC reactor. Two types of serpentine flow-through reactors are depicted in Figures 10.5(d) and 10.5(e), which

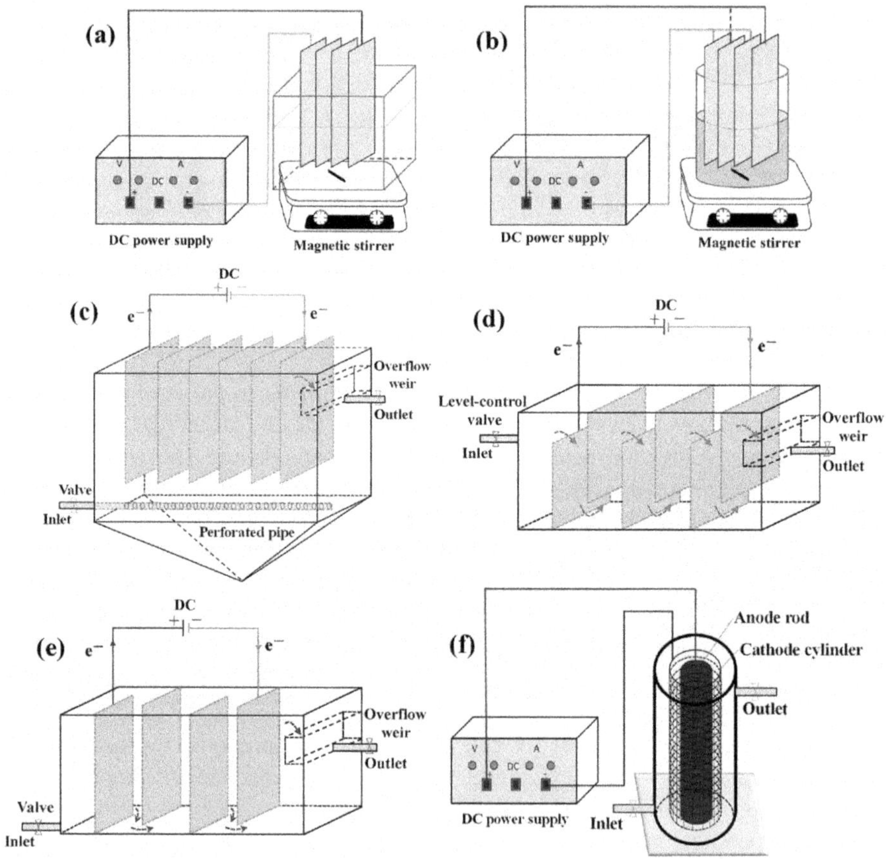

FIGURE 10.2 General EC reactor designs: (a) rectangular batch reactor (bipolar), (b) cylindrical batch reactor (monopolar), (c) vertical flow reactor with a perforated inlet pipe, (d) vertically serpentine flow reactor with baffle plates, (e) horizontally serpentine flow reactor with baffle plates, and (f) column reactor with a cylindrical cathode (Reproduced with permission from Lu et al. (2021) @ Elsevier [1]).

individually reflect the vertical flow and horizontal flow modes. Baffle electrode plates are used in these two reactors to create a single channel, which increases flow rate and improves mixing efficiency. Additionally, they are thought to increase the mass transfer of the coagulant from the anode to the bulk by causing fluid turbulence in the reactors. A typical cylindrical column reactor is shown in Figure 10.2(f) [1]. The cathode is a hollow cylindrical Fe mesh that is symmetrically positioned around the anode, which is a mild steel rod at its center. Both batch and continuous flow modes can be used with a cylindrical column reactor. The electrode pattern and form in this reactor can be designed with flexibility. For example, the inner electrode frequently serves as the anode and can be swapped out for a metallic ball, rod, or scrap metal. The cathode is inserted into the outer layer, which can be changed out for a perforated cylinder or metal mesh. Crucially, among the many varieties of EC reactors, the tall vertical-plate reactor is a well-liked form that may be expanded in accordance with treatment load. It is advised to employ a single channel to boost the flow rate when the electrodes' surface passivation cannot be minimized (Figures 10.2(d) and 10.2(e)) [49]. Because the EC reactors listed above have only static electrodes, poor diffusion and mass transfer make electrode passivation more likely to occur. In recent years, there have also been proposals for EC reactors with vibrating electrodes or rotating electrodes to counteract these negative consequences. For example, it is thought that the rotation of the anodes in an EC reactor can increase the reactor's overall process efficiency. As a result, it is simple to fix some issues including dead zone creation, back mixing, and fluid channeling/short circuiting [50].

Damaraju et al. (2020) looked at the use of a continuous bipolar-mode EC system in the field and a higher S/V ratio of 76.5 (m^2/m^3) for the treatment of marigold wastewater [51]. Reduced energy consumption and shorter retention times were associated with increased treatment efficiency. This study demonstrated how, when used on a field scale, EC systems with high S/V ratios might save space and energy. The current concentration (I/V) is the primary characteristic when the electrode surface area is sufficiently high.

$$E_{electrode} = (Mt)/(ZF)\cdot(I/V) \quad (10.6)$$

where $E_{electrode}$ and V stand for the EC cell's working volume (m^3) and the theoretical concentration of discharged metal (g/m^3), respectively. The other criteria remain the same as those previously stated.

10.4.2 Production of Hydrogen Gas During Electrocoagulation

The hydrogen gas generated during the EC process is one of the most advantageous aspects of EC technology. Hydrogen is a high-energy fuel that can be converted into electrical energy, reducing the process's overall energy requirements. Equation (10.7) was used to determine the quantity of H2 gas produced by the EC unit as a by-product [52].

$$A_{H2}(\text{mol}) = \frac{jAtNH_2}{F} \quad (10.7)$$

Since 1.0 mol of H_2 gas has 0.244 MJ of energy, or 0.0678 kWh of power, Eq. (10.8) can be used to calculate the power contents (P) of the H_2 gas that was collected.

$$PH_2\left(\text{kWh}\right) = A_{H2} \times 0.0678 \tag{10.8}$$

The effective surface area (m^2) of the electrodes is denoted by A, the electrocoagulation time (s) is represented by t, the applied current density (Am^{-2}) is represented by j, and the Faraday constant (F) is equal to 96,485 C mol^{-1}.

Oladipo et al. (2022) reported that the amount of hydrogen gas generated after treating a 0.2 L TC solution alone was in the range of 22.2–13.99 mol m^{-3} for either the Fe or Al electrode; however, in the TC–Ni binary mixture, it was 27.2–40.8 mol m^{-3} [52]. Equation (10.8) states that for single and binary mixtures, the amount of electrical power produced by recycling the emitted hydrogen gas is 1.84–2.77 kWh m^{-3} and 1.51–0.949 kWh m^{-3}, respectively. The findings demonstrate that the hydrogen produced during the EC treatment of a solution containing a 1:1 TC–Ni binary mixture can supply 63.2% and 35% of the electrical energy used for the Fe and Al electrodes, respectively, while the solution containing TC alone generated approximately 48.9–53.6% of the electrical energy consumed. These results imply that a considerable amount of energy can be saved by reusing the H_2 gas that is generated from field-scale EC plants when actual hospital or pharmaceutical wastewater is being treated.

The electrochemical reactor's effluent was continually delivered to the bottom of the gas separation tank, which was initially filled with water, according to the work carried out by Phalakornkule et al. (2010) [53]. Together with the wastewater and light, distributed sludge, upward effluent, and gas bubbles also rose to the top. The gas and water separated in the gas separation tank. The gas exit was at the top of the tank, and the water outlet was close to the top. Within a liquid displacement gas meter, the gas was collected. High liquid flow rates caused only moderate amounts of solids to collect in the gas separation tank, while light and dispersed sludge flowed out with the liquid. The amount of hydrogen extracted during the electrocoagulation process is displayed in Table 10.1. The product of the hydrogen purity determined by GC and the gas volume determined using a liquid displacement gas meter was used to determine the amount of hydrogen.

TABLE 10.1
Comparison of experimental and theoretical hydrogen yields (Reproduced with permission from Phalakornkule et al. (2010) @ Elsevier [53]).

Current density (J, Am^{-2})/EC time (min)	Theoretical amount of H_2		Experimental H_2 volume ($\times 10^3$ m^3 under normal conditions)	H_2 yield (experimental/theoretical)
	Mole	($\times 10^3$ m^3 under normal conditions)		
30 Am^{-2}/5 min	0.0244	0.587	0.521	0.89
40 Am^{-2}/5 min	0.0325	0.783	0.733	0.94
30 Am^{-2}/10 min	0.0488	1.173	1.039	0.89
40 Am^{-2}/10 min	0.0651	1.565	1.550	0.99

TABLE 10.2
Comparison of theoretical energy yield from harvested hydrogen product and energy requirement of the electrocoagulation process (Reproduced with permission from Phalakornkule et al. (2010) @ Elsevier [53]).

Current density (J, Am^{-2})/EC time (min)	Theoretical energy yield		Experimental energy usage (kWh_e mm^{-3})	
	$kWhm^{-3}$	$kWh_e m^{-3}$	Reactive Blue 140	Direct Red 23
30 Am^{-2}/5 min	0.178	0.089	0.81	0.69
40 Am^{-2}/5 min	0.237	0.119	1.42	1.23
30 Am^{-2}/10 min	0.356	0.178	1.63	1.38
40 Am^{-2}/10 min	0.474	0.237	2.83	2.46

The gas used at startup was a blend of hydrogen and air. As a result of the produced hydrogen replacing the air in the system, the purity of hydrogen increased with the operation time. The ratio of the amount of hydrogen obtained from the experiment to the amount of hydrogen calculated theoretically using Eq. (7) is defined as the hydrogen yield. Table 10.2 presents a comparison between the electrical energy needs of EC and the theoretical hydrogen production from the harvested hydrogen product. The electrocoagulation procedure used to treat Reactive Blue 140 required more electrical energy than the process used to treat Direct Red 23, which increased the system's resistance. The findings indicate that, assuming a 50% efficiency in the conversion of hydrogen to electricity, the energy output of captured hydrogen can lower the electrical energy demand of the electrocoagulation process by 8.5% for Reactive Blue 140 and 13% for Direct Red 23. The current densities and electrocoagulation periods used to cure various dye types have been reported in several investigations. It has been observed that the estimated hydrogen produced is always of the same order of magnitude, ranging between 0.1 and 0.4 $kWhm^{-3}$, regardless of the variations in the chemical structures of different dyes and the selection of current densities, electrocoagulation periods, anode types, and electrode diameters.

10.4.3 Management of Electrocoagulated Sludge

Before sludge undergoes a treatment procedure like thickening or dewatering, it must be processed and stored. Larger factories often store sludge in facultative lagoons, aerobic digesters, or using other long-term detention methods. Pretreatment is often necessary for sludge before dewatering or thickening treatment. It includes degritting and grinding. Sludge grinding is the process of reducing large solid sludge particles to smaller ones. This strategy is employed to prevent problems with the performance of downstream activities [54]. When cleaning and maintaining a machine, inline grinders reduce downtime. Sludge cakes can be held to improve operational reliability, balance out flow irregularities, and allow accumulation while downstream operations are not taking place. Sludge cakes are stored in hoppers or bins. These are available in several capacities, spanning from several cubic meters to 380 m^3.

Sludge dewatering techniques that are now in use can yield cakes with a solid concentration of 15–40%. These cakes have a consistency that varies from custard to wet cardboard. Prior to the dewatering or thickening stages, all of the sludge needs to go through conditioning treatments. Sludge conditioning boosts the efficiency of subsequent operations by applying heat or chemical treatment. The advantage of polymers over inorganics is that they do not dramatically increase the amount of sludge production; an additional kilogram of sludge is formed when inorganic compounds are administered [55]. One disadvantage of polymers is their relatively high cost. Sludge conditioning is affected by several important factors. Among its constituents are flocculation, coagulation, and sludge handling. The main objective of sludge conditioning is to promote the aggregation of fine particulates by means of either organic polymer flocculation, inorganic chemical coagulation, or both. For conditioning, dosage is an essential design component [56].

The process of stabilizing the sludge solids by adding different chemicals to the sludge matrix is known as chemical stabilization. Chlorine and lime stabilization are two methods that are frequently employed. Lime must be added to the sludge in sufficient amounts for the mixture's pH to be raised to 12 or higher. This often improves the performance of the hoover filter, reduces the risk of bacteria and odor to almost nothing, and provides sufficient means of stabilizing the sludge before disposal. To begin with, the answer is almost quick. Second, there is very little reduction in volatile compounds in the sludge. But most conditioning involves substituting or supplementing the organic component with chlorine to create new physiologically inert molecules [57].

Conditioning is the process of treating sludge solids with chemicals or other means in order to prepare them for dewatering procedures. Sludge conditioning, also known as chemical conditioning, prepares the sludge for treatment with vacuum filters or centrifuges, which are more efficient and economical methods. Acid salts, such as ferric chloride, are dissolved in tanks lined with materials resistant to acidity, such as rubber. A thorough mixing of the coagulant and sludge is necessary for optimal conditioning. Using ferric chloride, either alone or in combination with lime, is the most widely used conditioning technique. Many people are starting to approve the use of polymers. While lime and ferric chloride are often applied in combination, it is also possible to administer either component alone [58]. Two major methods can be used to thermally treat sludge. One is the oxidation of sludge without a flame in moist air at temperatures between 450 and 550°F and pressures of approximately 1200 psig. The other process, called heat treatment, accomplishes the same tasks at temperatures between 350 and 400°F and pressures between 150 and 300 psig. The sludge is converted to an ash by wet air oxidation (WAO), and it becomes more readily dewaterable with heat treatment [59]. Furthermore, sludge treated with heat is less difficult to dewater than sludge treated with chemicals. Sadly, biological organisms' cell walls rupture as a result of the heat treatment process, releasing some bonded organic material and water with them. As a result, some organic material that has been converted into particles is released and additional fine particulate matter is produced. Heat treatment breaks down biological cells, converting the once-particle cells back into fine solids and water. This helps with dewatering, but handling the highly contaminated liquid that leaks out of the cells is another challenge.

Dewatering the sludge is also known as thickening the sludge. Both mechanical and thermal techniques can be used to achieve this. A few mechanical methods for dewatering sludge include centrifugation, drum filters, belt filter presses, and pressure filter presses. Similar to vacuum filtration, pressure filtering involves removing the liquid from the solids in sludge. Without a doubt, leaf filters are the most often used type of filter. Leaf filters work similar to vacuum filters in that they separate the liquid from the particles using a porous substance [60]. The solids assist the media's ability to separate solids from liquids by being pulled into the pores and accumulating on the surface. The power required to force the water through the medium is provided by sludge pumps. Before pressing, sludge is typically preconditioned with lime, ferric salts, aluminum chloride, and aluminum chlorohydrate. Ash precoating is widely and effectively applied. Press filters are said to require less chemical expenses than vacuum filters, which is their primary advantage. Leaf filters are employed in an attempt to quickly dewater sludge in a confined space. They do, however, function in batches and have high maintenance and operating costs, which sets them apart from traditional dewatering processes. Other designs for pressure filters are screw presses and hydraulic systems, which, although effective in dewatering sludges, have the major disadvantage of frequently requiring a thicker sludge input. It is possible to create sludge cakes with up to 75% solid content by pressure filtering [61].

Centrifuges are apparatuses that extract particles from liquid by means of sedimentation and centrifugal force. The rotating conical bowl of a typical gadget has a screw conveyor put inside of it. In comparison to the bowl, it rotates noticeably more slowly. Sludge comes out of the end of the feed tube, picks up speed, passes through the ports on the conveyor shaft, and spreads out to the edge of the bowl. Centrifugation has certain inherent advantages over vacuum filtering and other techniques for dewatering sludge [62]. It is entirely enclosed, very simple, adaptable, and fairly priced. It can be used without the need for chemical support. Industries have embraced centrifuges for their great efficiency in processing difficult-to-dewater sludges, low initial cost, and ease of usage. The centrifuges with solid bowls, horizontal, cylindrical, and conical shapes work well at dewatering waste sludge. Basket centrifuges are an efficient way to dewater sludge; however, the liquid's purity is not high. Disc-type machines have strong dewatering capacity but are not perfect for clarifying [63].

10.5 CONCLUSION

Recent years have seen major developments and breakthroughs in the electrocoagulation process, which have deepened our grasp of its working and expanded its uses. Research has focused on four distinct areas of interest within electrocoagulation: sinusoidal alternating current electrocoagulation, air cathode electrocoagulation processes, the effect of H_2O_2 on electrocoagulation, and aerated electrocoagulation systems. Every one of these domains offers distinct perspectives and prospects to enhance the efficacy and efficiency of water treatment. Aerated electrocoagulation systems, which incorporate air into the electrolyte solution, present a viable way to improve the removal of contaminants. The overall efficiency of the process can be increased by enhancing the mass transfer of pollutants to the electrodes through the creation of microbubbles, which is encouraged by this aeration. In order to enhance treatment performance, research in this field is concentrated on optimizing aeration factors such as flow rate, bubble size, and dispersion. Utilizing air cathodes to boost oxygen reduction reactions and increase electrocoagulation efficiency, air cathode electrocoagulation methods offer yet another opportunity for innovation. Researchers want to improve the stability, boost treatment rates, and lower energy consumption of electrocoagulation systems by implementing air cathodes. In order to further enhance performance, future research may investigate innovative cathode materials and electrode configurations. Because it can increase the overall pollutant oxidation and treatment efficiency, H_2O_2's impact on electrocoagulation has drawn attention. Researchers want to increase the production of hydroxyl radicals in the electrolyte solution by adding H_2O_2. These radicals can aid in the breakdown of stubborn contaminants. There is potential for improving water treatment capacities by examining the synergistic effects of H_2O_2 and electrocoagulation, especially when it comes to handling complex wastewater streams and new contaminants. An innovative strategy to maximize electrode passivation and maximize treatment efficiency is sinusoidal alternating current electrocoagulation. Researchers aim at addressing problems like electrode fouling and scale building that can impair the efficiency of conventional direct current electrocoagulation systems by utilizing alternating current with a sinusoidal waveform. In order to enhance therapy effectiveness, future study may concentrate on waveform parameter optimization and investigation of the underlying electrochemical mechanisms.

Future developments in electrocoagulation technology are expected to be shaped by a number of significant research trends. These include improvements in reactor design and important scaling characteristics that make it easier to move from small-scale research in the lab to large-scale commercial applications. Furthermore, the formation of hydrogen gas during electrocoagulation is gaining attention due to its possible implications for resource recovery and the production of renewable energy. Lastly, the management of electrocoagulated sludge continues to be a significant challenge, requiring the development of economical and environmentally friendly methods for stabilization, dewatering, and disposal of sludge. In summary, there is potential for improving sustainable water treatment solutions and tackling issues with water quality through continued research into electrocoagulation processes and their varied uses. Through sustained innovation and interdisciplinary collaboration, scientists may fully realize the potential of electrocoagulation technology and establish the foundation for a future that is more robust and resource-efficient.

REFERENCES

[1] J. Lu, P. Zhang, J. Li, Electrocoagulation technology for water purification: An update review on reactor design and some newly concerned pollutants removal, *J. Environ. Manage.* 296 (2021) 113259. https://doi.org/10.1016/j.jenvman.2021.113259.

[2] P.P. Das, A. Anweshan, M.K. Purkait, Treatment of cold rolling mill (CRM) effluent of steel industry, *Sep. Purif. Technol.* 274 (2021) 119083. https://doi.org/10.1016/j.seppur.2021.119083.

[3] P.P. Das, P. Mondal, Anweshan, A. Sinha, P. Biswas, S. Sarkar, M.K. Purkait, Treatment of steel plant generated biological oxidation treated (BOT) wastewater by hybrid process, *Sep. Purif. Technol.* 258 (2021) 118013. https://doi.org/10.1016/j.seppur.2020.118013.

[4] P.P. Das, A. Anweshan, P. Mondal, A. Sinha, P. Biswas, S. Sarkar, M.K. Purkait, Integrated ozonation assisted electrocoagulation process for the removal of cyanide from steel industry wastewater, *Chemosphere* 263 (2021) 128370. https://doi.org/10.1016/j.chemosphere.2020.128370.
[5] P.P. Das, M. Sharma, M.K. Purkait, Recent progress on electrocoagulation process for wastewater treatment: A review, *Sep. Purif. Technol.* 292 (2022) 121058. https://doi.org/10.1016/j.seppur.2022.121058.
[6] P.P. Das, A.D. Sontakke, M.K. Purkait, Electrocoagulation process for wastewater treatment: Applications, challenges, and prospects, in: M.P. Shah (Ed.), *Development in Wastewater Treatment Research and Processes*, Elsevier, 2023: pp. 23–48. https://doi.org/10.1016/B978-0-323-95684-0.00015-4.
[7] P.P. Das, S. Dhara, M.K. Purkait, Ozone-based oxidation processes for the removal of pharmaceutical products from wastewater, in: M.P. Shah, P. Ghosh (Eds.), *Development in Wastewater Treatment Research and Processes*, Elsevier, 2024: pp. 287–308. https://doi.org/10.1016/B978-0-443-19207-4.00003-3.
[8] P.P. Das, S. Dhara, M.K. Purkait, Hybrid electrocoagulation and ozonation techniques for industrial wastewater treatment, in: M.P. Shah (Ed.), *Sustainable Industrial Wastewater Treatment and Pollution Control*, Springer Nature Singapore, Singapore, 2023: pp. 107–128. https://doi.org/10.1007/978-981-99-2560-5_6.
[9] P.P. Das, A.D. Sontakke, N.S. Samanta, M.K. Purkait, Emerging contaminants in wastewater: Eco-toxicity and sustainability assessment, in: M.P. Shah (Ed.), *Industrial Wastewater Reuse: Applications, Prospects and Challenges*, Springer Nature Singapore, Singapore, 2023: pp. 63–87. https://doi.org/10.1007/978-981-99-2489-9_4.
[10] P.V. Nidheesh, A.A. Oladipo, N.G. Yasri, A.R. Laiju, V.R.S. Cheela, A. Thiam, Y.G. Asfaha, S. Kanmani, E. (Ted) P.L. Roberts, Emerging applications, reactor design and recent advances of electrocoagulation process, *Process Saf. Environ. Prot.* 166 (2022) 600–616. https://doi.org/10.1016/j.psep.2022.08.051.
[11] M. Sharma, P.P. Das, A. Chakraborty, M.K. Purkait, Clean energy from salinity gradients using pressure retarded osmosis and reverse electrodialysis: A review, *Sustain. Energy Technol. Assessments* 49 (2022) 101687. https://doi.org/10.1016/j.seta.2021.101687.
[12] L.C. Roberts, S.J. Hug, T. Ruettimann, M.M. Billah, A.W. Khan, M.T. Rahman, Arsenic removal with Iron(II) and Iron(III) in waters with high silicate and phosphate concentrations, *Environ. Sci. Technol.* 38 (2004) 307–315. https://doi.org/10.1021/es0343205.
[13] M. Sharma, P.P. Das, T. Sood, A. Chakraborty, M.K. Purkait, Reduced graphene oxide incorporated polyvinylidene fluoride/cellulose acetate proton exchange membrane for energy extraction using microbial fuel cells, *J. Electroanal. Chem.* 907 (2022) 115890. https://doi.org/10.1016/j.jelechem.2021.115890.
[14] M. Sharma, P.P. Das, T. Sood, A. Chakraborty, M.K. Purkait, Ameliorated polyvinylidene fluoride based proton exchange membrane impregnated with graphene oxide, and cellulose acetate obtained from sugarcane bagasse for application in microbial fuel cell, *J. Environ. Chem. Eng.* 9 (2021) 106681. https://doi.org/10.1016/j.jece.2021.106681.
[15] D. Syam Babu, P. V Nidheesh, M. Suresh Kumar, Arsenite removal from aqueous solution by aerated iron electrocoagulation process, *Sep. Sci. Technol.* 56 (2021) 184–193. https://doi.org/10.1080/01496395.2019.1708932.
[16] D. Syam Babu, K. Vijay, P.V. Nidheesh, M. Suresh Kumar, Performance of continuous aerated iron electrocoagulation process for arsenite removal from simulated groundwater and management of arsenic-iron sludge, *Sustain. Energy Technol. Assessments* 47 (2021) 101476. https://doi.org/10.1016/j.seta.2021.101476.
[17] C.M. van Genuchten, S.R.S. Bandaru, E. Surorova, S.E. Amrose, A.J. Gadgil, J. Peña, Formation of macroscopic surface layers on Fe(0) electrocoagulation electrodes during an extended field trial of arsenic treatment, *Chemosphere* 153 (2016) 270–279. https://doi.org/10.1016/j.chemosphere.2016.03.027.

[18] H.A. Maitlo, J. Lee, J.Y. Park, J.-C. Kim, K.-H. Kim, J.H. Kim, An energy-efficient air-breathing cathode electrocoagulation approach for the treatment of arsenite in aquatic systems, *J. Ind. Eng. Chem.* 73 (2019) 205–213. https://doi.org/10.1016/j.jiec.2019.01.026.
[19] N.S. Samanta, P.P. Das, P. Mondal, U. Bora, M.K. Purkait, Physico-chemical and adsorption study of hydrothermally treated zeolite A and FAU-type zeolite X prepared from LD (Linz—Donawitz) slag of the steel industry, *Int. J. Environ. Anal. Chem.* (n.d.) 1–23. https://doi.org/10.1080/03067319.2022.2079082.
[20] N.S. Samanta, P.P. Das, P. Mondal, M. Changmai, M.K. Purkait, Critical review on the synthesis and advancement of industrial and biomass waste-based zeolites and their applications in gas adsorption and biomedical studies, *J. Indian Chem. Soc.* 99 (2022) 100761. https://doi.org/10.1016/j.jics.2022.100761.
[21] Y. Si, G. Li, Y. Wu, H. Zhang, Y. Yuan, H. Zhang, B. Liu, F. Zhang, Tradeoff between groundwater arsenite removal efficiency and current production in the self-powered air cathode electrocoagulation with different oxygen reduction pathways, *J. Hazard. Mater.* 357 (2018) 138–145. https://doi.org/10.1016/j.jhazmat.2018.05.048.
[22] K. Xu, J. Lu, I.D. Tegladza, Q. Xu, Z. Yang, G. Lv, Combined metal/air fuel cell and electrocoagulation process: Energy generation, flocs production and pollutant removal, *Chemosphere* 255 (2020) 126925. https://doi.org/10.1016/j.chemosphere.2020.126925.
[23] T. Mu, M. Park, K.-Y. Kim, Energy-efficient removal of PFOA and PFOS in water using electrocoagulation with an air-cathode, *Chemosphere* 281 (2021) 130956. https://doi.org/10.1016/j.chemosphere.2021.130956.
[24] N. Shekhar Samanta, P.P. Das, S. Dhara, M.K. Purkait, An Overview of Precious Metal Recovery from Steel Industry Slag: Recovery Strategy and Utilization, *Ind. Eng. Chem. Res.* 62 (2023) 9006–9031. https://doi.org/10.1021/acs.iecr.3c00604.
[25] J.H. Kim, B. min An, D.H. Lim, J.Y. Park, Electricity production and phosphorous recovery as struvite from synthetic wastewater using magnesium-air fuel cell electrocoagulation, *Water Res.* 132 (2018) 200–210. https://doi.org/10.1016/j.watres.2018.01.003.
[26] A. Almukdad, M.A. Hafiz, A.T. Yasir, R. Alfahel, A.H. Hawari, Unlocking the application potential of electrocoagulation process through hybrid processes, *J. Water Process Eng.* 40 (2021) 101956. https://doi.org/10.1016/j.jwpe.2021.101956.
[27] S. Dhara, N. Shekhar Samanta, P.P. Das, R.V.S. Uppaluri, M.K. Purkait, Ravenna grass-extracted alkaline lignin-based polysulfone mixed matrix membrane (MMM) for Aqueous Cr(VI) removal, *ACS Appl. Polym. Mater.* 5 (2023) 6399–6411. https://doi.org/10.1021/acsapm.3c00999.
[28] F. Haber, J. Weiss, J.O. Seph, W. Eiss, The catalytic decomposition o f hydrogen peroxide by iron salts, *R. Soc.* 147 (1934) 332–351. https://doi.org/10.1098/rspa.1934.0221.
[29] G. Roa-Morales, E. Campos-Medina, J. Aguilera-Cotero, B. Bilyeu, C. Barrera-Díaz, Aluminum electrocoagulation with peroxide applied to wastewater from pasta and cookie processing, *Sep. Purif. Technol.* 54 (2007) 124–129. https://doi.org/10.1016/j.seppur.2006.08.025.
[30] S. Vasudevan, J. Lakshmi, Effect of alternating and direct current in an electrocoagulation process on the removal of cadmium from water, *Water Sci. Technol.* 65 (2012) 353–360. https://doi.org/10.2166/wst.2012.859.
[31] T. Xu, Y. Zhou, X. Lei, B. Hu, H. Chen, G. Yu, Study on highly efficient Cr(VI) removal from wastewater by sinusoidal alternating current coagulation, *J. Environ. Manage.* 249 (2019) 109322. https://doi.org/10.1016/j.jenvman.2019.109322.
[32] M. Changmai, P.P. Das, P. Mondal, M. Pasawan, A. Sinha, P. Biswas, S. Sarkar, M.K. Purkait, Hybrid electrocoagulation—microfiltration technique for treatment of nanofiltration rejected steel industry effluent, *Int. J. Environ. Anal. Chem.* 102 (2022) 62–83. https://doi.org/10.1080/03067319.2020.1715381.

[33] J. Abdollahi, M.R. Alavi Moghaddam, S. Habibzadeh, The role of the current waveform in mitigating passivation and enhancing electrocoagulation performance: A critical review, *Chemosphere* 312 (2023) 137212. https://doi.org/10.1016/j.chemosphere.2022.137212.
[34] H.J. Mansoorian, A.H. Mahvi, A.J. Jafari, Removal of lead and zinc from battery industry wastewater using electrocoagulation process: Influence of direct and alternating current by using iron and stainless steel rod electrodes, *Sep. Purif. Technol.* 135 (2014) 165–175. https://doi.org/10.1016/j.seppur.2014.08.012.
[35] A.D. Sontakke, P.P. Das, P. Mondal, M.K. Purkait, Thin-film composite nanofiltration hollow fiber membranes toward textile industry effluent treatment and environmental remediation applications: Review, *Emergent Mater.* (2021). https://doi.org/10.1007/s42247-021-00261-y.
[36] V.M. García-Orozco, I. Linares-Hernández, R. Natividad, P. Balderas-Hernández, C. Alanis-Ramírez, C.E. Barrera-Díaz, G. Roa-Morales, Solar-photovoltaic electrocoagulation of wastewater from a chocolate manufacturing industry: Anodic material effect (aluminium, copper and zinc) and life cycle assessment, *J. Environ. Chem. Eng.* 10 (2022) 107969. https://doi.org/10.1016/j.jece.2022.107969.
[37] Z. Mohamad, A.A. Razak, S. Krishnan, L. Singh, A.W. Zularisam, M. Nasrullah, Treatment of palm oil mill effluent using electrocoagulation powered by direct photovoltaic solar system, *Chem. Eng. Res. Des.* 177 (2022) 578–582. https://doi.org/10.1016/j.cherd.2021.11.019.
[38] Y. Dong, Y. Qu, C. Li, X. Han, J.J. Ambuchi, J. Liu, Y. Yu, Y. Feng, Simultaneous algae-polluted water treatment and electricity generation using a biocathode-coupled electrocoagulation cell (bio-ECC), *J. Hazard. Mater.* 340 (2017) 104–112. https://doi.org/10.1016/j.jhazmat.2017.06.055.
[39] M. Kobya, M. Bayramoglu, M. Eyvaz, Techno-economical evaluation of electrocoagulation for the textile wastewater using different electrode connections, *J. Hazard. Mater.* 148 (2007) 311–318. https://doi.org/10.1016/j.jhazmat.2007.02.036.
[40] J.N. Hakizimana, B. Gourich, M. Chafi, Y. Stiriba, C. Vial, P. Drogui, J. Naja, Electrocoagulation process in water treatment: A review of electrocoagulation modeling approaches, *Desalination* 404 (2017) 1–21. https://doi.org/10.1016/j.desal.2016.10.011.
[41] K.S. Hashim, P. Kot, S.L. Zubaidi, R. Alwash, R. Al Khaddar, A. Shaw, D. Al-Jumeily, M.H. Aljefery, Energy efficient electrocoagulation using baffle-plates electrodes for efficient Escherichia coli removal from wastewater, *J. Water Process Eng.* 33 (2020) 101079. https://doi.org/10.1016/j.jwpe.2019.101079.
[42] V. Khandegar, A.K. Saroha, Effect of Electrode Shape and Current Source on Performance of Electrocoagulation, *J. Hazardous Toxic Radioact. Waste* 20 (2016) 2–5. https://doi.org/10.1061/(asce)hz.2153-5515.0000278.
[43] M.H. Ibrahim, D.T. Moussa, M.H. El-Naas, M.S. Nasser, A perforated electrode design for passivation reduction during the electrochemical treatment of produced water, *J. Water Process Eng.* 33 (2020) 101091. https://doi.org/10.1016/j.jwpe.2019.101091.
[44] B. Abdulhadi, P. Kot, K. Hashim, A. Shaw, M. Muradov, R. Al-Khaddar, Continuous-flow electrocoagulation (EC) process for iron removal from water: Experimental, statistical and economic study, *Sci. Total Environ.* 760 (2021) 143417. https://doi.org/10.1016/j.scitotenv.2020.143417.
[45] U. Tezcan Un, E. Aytac, Electrocoagulation in a packed bed reactor-complete treatment of color and cod from real textile wastewater, *J. Environ. Manage.* 123 (2013) 113–119. https://doi.org/10.1016/j.jenvman.2013.03.016.
[46] U. Tezcan Un, A. Kandemir, N. Erginel, S.E. Ocal, Continuous electrocoagulation of cheese whey wastewater: An application of response surface methodology, *J. Environ. Manage.* 146 (2014) 245–250. https://doi.org/10.1016/j.jenvman.2014.08.006.

[47] L. Xu, G. Cao, X. Xu, C. He, Y. Wang, Q. Huang, M. Yang, Sulfite assisted rotating disc electrocoagulation on cadmium removal: Parameter optimization and response surface methodology, *Sep. Purif. Technol.* 195 (2018) 121–129. https://doi.org/10.1016/j.seppur.2017.12.010.

[48] L.F. Castañeda, O. Coreño, J.L. Nava, G. Carreño, Removal of fluoride and hydrated silica from underground water by electrocoagulation in a flow channel reactor, *Chemosphere* 244 (2020) 125417. https://doi.org/10.1016/j.chemosphere.2019.125417.

[49] N.S. Graça, A.M. Ribeiro, A.E. Rodrigues, Removal of fluoride from water by a continuous electrocoagulation process, *Ind. Eng. Chem. Res.* 58 (2019) 5314–5321. https://doi.org/10.1021/acs.iecr.9b00019.

[50] O.J. Flores, J.L. Nava, G. Carreño, E. Elorza, F. Martínez, Arsenic removal from groundwater by electrocoagulation in a pre-pilot-scale continuous filter press reactor, *Chem. Eng. Sci.* 97 (2013) 1–6. https://doi.org/10.1016/j.ces.2013.04.029.

[51] M. Damaraju, D. Bhattacharyya, T.K. Panda, K.K. Kurilla, Marigold wastewater treatment in a lab-scale and a field-scale continuous bipolar-mode electrocoagulation system, *J. Clean. Prod.* 245 (2020) 118693. https://doi.org/10.1016/j.jclepro.2019.118693.

[52] A.A. Oladipo, F.S. Mustafa, O.N. Ezugwu, M. Gazi, Efficient removal of antibiotic in single and binary mixture of nickel by electrocoagulation process: Hydrogen generation and cost analysis, *Chemosphere* 300 (2022) 134532. https://doi.org/10.1016/j.chemosphere.2022.134532.

[53] C. Phalakornkule, P. Sukkasem, C. Mutchimsattha, Hydrogen recovery from the electrocoagulation treatment of dye-containing wastewater, *Int. J. Hydrogen Energy* 35 (2010) 10934–10943. https://doi.org/10.1016/j.ijhydene.2010.06.100.

[54] Z. Košnář, F. Mercl, L. Pierdonà, A.D. Chane, P. Míchal, P. Tlustoš, Concentration of the main persistent organic pollutants in sewage sludge in relation to wastewater treatment plant parameters and sludge stabilisation, *Environ. Pollut.* 333 (2023) 122060. https://doi.org/10.1016/j.envpol.2023.122060.

[55] W. Wu, X. Li, B. Zhou, Z. Wang, Impacts of floc breakage on dewaterability of chemically conditioned sludges and implications on practical conditioning strategies, *Chem. Eng. J.* 459 (2023) 141626. https://doi.org/10.1016/j.cej.2023.141626.

[56] W. Lin, J. Zeng, R. Zhang, X. He, J. Nan, G. Li, J. Ma, H.H. Ngo, A. Ding, Selection of metal ions in different valence states on sludge conditioning: Analysis of hydrophobicity and evaluation of resource recovery capacity, *J. Water Process Eng.* 50 (2022) 103297. https://doi.org/10.1016/j.jwpe.2022.103297.

[57] S. Pottipati, A.S. Kalamdhad, Thermophilic-mesophilic biodegradation: An optimized dual-stage biodegradation technique for expeditious stabilization of sewage sludge, *J. Environ. Manage.* 323 (2022) 116189. https://doi.org/10.1016/j.jenvman.2022.116189.

[58] X. Fan, Y. Wang, D. Zhang, S. Zhang, C. Liu, M. Liu, A comprehensive assessment on sludge conditioning by pyrite acid eluent-activated peroxymonosulfate based on dewaterability, heavy metals risk and ore recovery, *Waste Manag.* 170 (2023) 82–92. https://doi.org/10.1016/j.wasman.2023.08.001.

[59] Y. Wang, M.X. Gu, D. Ge, Y. Dong, L. Bai, Y. Han, N. Zhu, Polyhexamethylene biguanidine used as a new type sewage sludge conditioning agent: Effect on sludge dewaterability and mechanism, *J. Environ. Manage.* 315 (2022) 115146. https://doi.org/10.1016/j.jenvman.2022.115146.

[60] B. Wu, X. Dai, X. Chai, Critical review on dewatering of sewage sludge: Influential mechanism, conditioning technologies and implications to sludge re-utilizations, *Water Res.* 180 (2020) 115912. https://doi.org/10.1016/j.watres.2020.115912.

[61] X. Zhang, P. Ye, Y. Wu, Enhanced technology for sewage sludge advanced dewatering from an engineering practice perspective: A review, *J. Environ. Manage.* 321 (2022) 115938. https://doi.org/10.1016/j.jenvman.2022.115938.

[62] B. Cao, T. Zhang, W. Zhang, D. Wang, Enhanced technology based for sewage sludge deep dewatering: A critical review, *Water Res.* 189 (2021) 116650. https://doi.org/10.1016/j.watres.2020.116650.
[63] B. Cao, W. Zhang, Q. Wang, Y. Huang, C. Meng, D. Wang, Wastewater sludge dewaterability enhancement using hydroxyl aluminum conditioning: Role of aluminum speciation, *Water Res.* 105 (2016) 615–624. https://doi.org/10.1016/j.watres.2016.09.016.

Index

W

Y

Z

For Product Safety Concerns and Information please contact our EU
representative GPSR@taylorandfrancis.com
Taylor & Francis Verlag GmbH, Kaufingerstraße 24, 80331 München, Germany

www.ingramcontent.com/pod-product-compliance
Ingram Content Group UK Ltd.
Pitfield, Milton Keynes, MK11 3LW, UK
UKHW020204200926
475702UK00004B/604

* 9 7 8 1 0 3 2 7 9 6 5 0 5 *